Local and Urban Governance

Series Editor
Carlos Nunes Silva , Institute of Geography and Spatial Planning, University of Lisbon, Lisbon, Portugal

This series contains research studies with policy relevance in the field of sub-national territorial governance, at the micro, local and regional levels, as well as on its connections with national and supranational tiers. The series is multidisciplinary and brings together innovative research from different areas within the Social Sciences and Humanities. The series is open for theoretical, methodological and empirical ground breaking contributions. Books included in this series explore the new modes of territorial governance, new perspectives and new research methodologies. The aim is to present advances in Governance Studies to scholars and researchers in universities and research organizations, and to policy makers worldwide. The series includes monographs, edited volumes and textbooks. Book proposals and final manuscripts are peer-reviewed.

The areas covered in the series include but are not limited to the following subjects:

- Local and regional government
- Urban and metropolitan governance
- Multi-level territorial governance
- Post-colonial local governance
- Municipal merger reforms
- Inter-municipal cooperation
- Decentralized cooperation
- Governance of spatial planning
- Strategic spatial planning
- Citizen participation in local policies
- Local governance, spatial justice and the right to the city
- Local public services
- Local economic development policies
- Entrepreneurialism and municipal public enterprises
- Local government finance
- Local government and sustainable development
- Anthropocene and green local governance
- Climate change and local governance
- Smart local governance

The series is intended for geographers, planners, political scientists, sociologists, lawyers, historians, urban anthropologists and economists. **Local and Urban Governance—now indexed in Scopus**

Khadija Darmame • Eric Ross
Editors

Local Governance
and Development in Africa
and the Middle East

 Springer

Editors
Khadija Darmame
Al Akhawayn University in Ifrane
Ifrane, Morocco

Eric Ross
Al Akhawayn University in Ifrane
Ifrane, Morocco

ISSN 2524-5449 ISSN 2524-5457 (electronic)
Local and Urban Governance
ISBN 978-3-031-60659-5 ISBN 978-3-031-60657-1 (eBook)
https://doi.org/10.1007/978-3-031-60657-1

This Springer imprint is published by the registered company Springer Nature Switzerland AG
The registered company address is: Gewerbestrasse 11, 6330 Cham, Switzerland

If disposing of this product, please recycle the paper.

Map locating the local governance and development issues discussed in this volume

Legend:
- ⑩ Chapters
- ■ Cities
- ✕ Resource extraction communities
- O Resource management communities
- ✦ Spring

Preface

This volume gestated during the COVID 19 pandemic. The co-editors were eager to contribute to the International Geography Union's concerted inquiry into the geography of governance by organizing a conference on the theme in Morocco that would cover African and Middle Eastern experiences of local governance and development. By the time the conference was held, in February 2022, much of the world was still abiding by various degrees of stay-at-home isolation. Consequently, participants contributed via video-conference, which limited the intensity of interaction that usually characterizes good scholarly meetings. To compensate, we sought additional contributions from researchers who had not attended the on-line event, ending up with the collection of 14 chapters we have brought together here.

Ifrane, Morocco

Khadija Darmame
Eric Ross

Acknowledgements

Our effort to publish this work was greatly facilitated by Carlos Nunes Silva, Professor at the Institute of Geography and Spatial Planning, University of Lisbon. His unflinching encouragement to persevere in traducing disparate conference papers into a coherent volume was exactly what we needed to hear. Our thanks also go out to Robert Home, Professor Emeritus of Land Management at Anglia Ruskin University. Home regularly enquired about the state of the manuscript, which helped keep it at the top of our "to do" list as other, less fulfilling, professional obligations slowly piled up. He also volunteered to take on a number of tasks associated with this publication. It was comforting to know someone was looking out for us. Finally, our own institution, Al Akhawayn University's School of Humanities and Social Sciences, provided funding to have contributions translated into English.

Contents

Part II   Urban Planning and Development

Part III   Public Services, Housing and Public Space

Part IV   Resource Management

Contributors

Atman Aoui President of the Moroccan Association for the Promotion of Mediation, Tinghir Province, Morocco

Ahmed Haj Asaad Geo Expertise, Geneve, Switzerland

Esther Yeboah Danso-Wiredu Department of Geography Education, University of Education, Winneba, Ghana

Khadija Darmame Al Akhawayn University in Ifrane, Ifrane, Morocco

Christopher Dick-Sagoe University of Botswana, Gaborone, Botswana

Moulay Ahmed el Amrani Civil Society Activist, Tinghir Province, Morocco

Mohamed El-Mensi Governance and Local Development Consultant, Senior partner - Local Development International, New York City, NY, USA

El-Hassan Farhat Polydisciplinary Faculty of Safi, Cadi Ayyad University, Marrakech, Morocco

Robert Home Department of Business and Law, Anglia Ruskin University, Cambridge, UK

Samira Idllalène Polydisciplinary Faculty, Law Studies Department, Cadi Ayyad University, Safi, Morocco
Faculty of Legal, Economic and Social Sciences, Laboratoire de recherche sur la coopération internationale pour le développement (LRCID), Marrakech, Morocco

Abdelkader Kaioua School of Letters and Social Sciences, Hassan II University, Casablanca, Morocco

Fuad Malkawi Senior Urban Specialist at the World Bank, Washington, DC, USA

Zeina Moneer Research Institute for a Sustainable Environment, American University in Cairo, New Cairo, Egypt

Safaa Monqid Université Paris 3, Sorbonne nouvelle, Paris, France

Carol Chi Ngang National University of Lesotho, Roma, Lesotho
Research Fellow, Free State Centre for Human Rights, University of the Free State, Bloemfontein, South Africa

André Akono Olinga Faculty of Legal and Political Sciences, University of Yaoundé 2, Soa, Cameroon

Margot Petitpierre General Manager for Asia, Insuco, Phnom Penh, Cambodia

Pascal Rey Institute for Social Research in Africa (IFSRA), Ouagadougou, Burkina Faso

Karen Rignall Community and Leadership Development Department, College of Agriculture, Food and Environment, University of Kentuky, Lexington, KY, USA

Eric Ross Al Akhawayn University in Ifrane, Ifrane, Morocco

List of Figures

List of Tables

Chapter 1
Introduction: Territorial Governance for Equitable and Sustainable Development: Examples from Africa and the Middle East

Khadija Darmame and Eric Ross

1.1 Introduction

This edited volume attempts to set out a comprehensive framework for understanding the challenges posed by territorial governance and local development practices across Africa and the Middle East. As a field for exercising politico-administrative power, territorial management is central to public policies at the local level. New modes of governance require local actors and policymakers to tailor their strategies based on a spatial approach. This territorialization of policies necessitates continuous coordination and consultation at multiple levels. The goal is to move from public policy to public action by adapting national orientations and sectoral policies to the specificities and potential of every locality. Territorial governance can be effective by making the territory more sustainable, competitive, and equitable by addressing regional socio-economic disparities and unequal distribution of resources. The volume is the outcome of an on-line conference on local governance and development held at Al Akhawayn University in Ifrane, Morocco, on 11–13 February 2022. The conference was organized in collaboration with the International Geography Union's Commission on Geography of Governance, and the African Local Governments Academy (ALGA) of United Cities and Local Governments of Africa (UCLG Africa). The 14 contributions introduce readers to a plethora of themes and issues raised by interdisciplinary scholars and practitioners who research and work in these regions. Most of the contributors are geographers of one stripe or another and together they provide a collective exploration, a mosaic of insights and experiences, best practices and visions.

K. Darmame (✉) · E. Ross
Al Akhawayn University in Ifrane, Ifrane, Morocco
e-mail: K.Darmame@aui.ma; E.Ross@aui.ma

© The Author(s), under exclusive license to Springer Nature Switzerland AG 2024
K. Darmame, E. Ross (eds.), *Local Governance and Development in Africa and the Middle East*, Local and Urban Governance,
https://doi.org/10.1007/978-3-031-60657-1_1

1

Additionally, this volume surveys Africa and the Middle East, two world regions which are usually presented separately in the Development Studies literature (Aïdi et al. 2020). This artificial separation is hardly warranted as the two regions grapple with very similar. Following Middell (2017), we believe that a trans-regional perspective is crucial to understanding the aftermath of several major crises: anti-government uprisings, the fall and rise again of authoritarian regimes, multiple health crises (e.g. Ebola and COVID 19), political transition and instability, food and energy crises, chaotic climate disasters, rapid uncontrolled urbanization, forced migration, hegemonic neoliberalism, and the patent failure of public polices ostensibly meant to reduce poverty and enhance social inclusion.

We delve deeply into these overlapping political, economic, social and environmental problems by analyzing them through spatial, territorial and scalar lenses. We start with the assumption that the problems and dysfunctions we discuss are *emplaced*, whereas the policies and strategies adopted to resolve them are too often place-less. This issue reveals the urgent need to foster shared reflections on how to implement collective governance and a human rights-based approach while developing territories. Such reflections should not only be driven by global forces or national policies but, also, by place-specific local policies and community strategies. In other words, localities matter to development outcomes, whether those localities are remote villages or sprawling metropolitan regions.

This volume presents insights and innovative approaches to fostering inclusive governance and sustainable development which have been tried by a variety of actors across Africa and the Middle East. By surveying these actions and struggles, their successes and their failures, we aim to demonstrate the power of the all-to-often forgotten local scale.

1.2   Linking Governance to Local Development

Why does governance matter for development? What are the connections between governance, territory, and development? Why is the local scale so crucial to building an effective and inclusive territorial governance system? To address these questions, it is necessary to first define concepts linked to territorial governance.

The concept of governance, seen as a broader term than government, concerns the traditions and institutions by which authority is exercised, and the process of interaction and decision-making between different types of actors. It includes the capacity of governments to formulate and implement effective policies and the attitude of citizens toward the institutions that govern them (Houston and Harding 2013). The concept took another turn in the 1990s, especially in developing countries, with an emphasis on local development through the transfer of responsibilities and duties from central structures to local governments. The aim is to deal with social, economic, and territorial development by devolving powers within the framework of decentralization, making local governance a powerful pillar of political systems. Therefore, meeting the expectations of citizens and overcoming devclopmental challenges requires empowering and strengthening local government performance. Local governments are the closest institutions to communities, more responsive to local needs, and make better use of resources by directing them towards providing basic social services (United Nation Development Program 2010). This cannot be done without real reform of administrative structures and legal frameworks, and enhancement of the capacities of local civil servants.

1.2.1 Development Agendas and Governance Models: One Size Does Not Fit All

Development is currently the main agenda for organizations and institutions of both national and international entities. It is considered the most efficient method of ensuring the well-being of citizens across the globe. However, due to the complexity of variated development challenges, it is also necessary to create unique action plans able to resolve each one of them. A governance model is the path that entities usually follow to implement a development strategy to meet these challenges. The model determines and controls the institutions and policies holistically. Governance is multi-scalar and is setup based on the objectives determined by decision-makers. Hooghe and Marks (2016) explored in their research the focus of each type of governance; "Type I governance, they describe, is designed around human (usually territorial) communities, while Type II is designed around particular tasks or policy problems." (Stead 2014).

According to the Sustainable Development Goals Fund (SDGF), development can only be achieved by developing the capacity of human beings and taking them as the focus of agendas (United Nations Industrial Development Organization 2016). This shift imposes a change in the approach and a redefinition of governance and management approaches. In order to develop individuals, it is considered necessary to create setups consisting of adequate environments that allow such individuals to focus on their own social or financial development.

1.2.2 Territorial Governance in the Making[1]

In his work on territories and governance processes, Torre (2010) states that territories must deal with three major changes that impacted the governance of their local structure:

1. The growing number of local actors and the complexity of their involvement, mainly in peri-urban and rural areas,

---

[1]For more insights on this point, see Davoudi et al. (2008).

2. Strong involvement of local communities in decision-making processes and territorial management (such as pressure or action groups like CSOs, or more or less formal lobbies),
3. A complicated mosaic of actors involved at various levels with specific sets of decisions and regulations.

The path to resolving the challenges of development is territorial governance, which is a relatively new and complex concept best defined as a set of mechanisms, structures and processes bringing together the place-based approach and multi-level governance within a given territory (European Union 2020). Territorial management, as a field of exercise of politico-administrative power at the local level, is central to public policies (Oosterhof 2018). It covers how power, resources, and responsibilities are distributed among different stakeholders, including governmental institutions, Civil Society Organizations (CSOs) and Community Based Organizations (CBOs). Therefore, the governance of territories is not limited to an idyllic vision of economic and social relations, i.e. forms of cooperation and common structures (Torre and Traversac 2011), but rests on the interactions of heterogeneous groups of actors at various levels of government (central, regional, and local) in making decisions, formulating and assessing local development plans, implementing programs and projects, allocating and administrating resources, and enhancing social, economic cultural and environmental wellbeing. The ultimate goal is to contribute to the development of broad consultation mechanisms on future paths of development and cooperation (Leloup et al. 2005; Pasquier et al. 2007; Torre 2011).

Hence, this framework can be used, in the first instance, to understand the environment of individuals, to pinpoint specific needs, and to develop programs that address them. Second, territorial governance can also be used in determining an agenda, and finally in the project implementation process. Accordingly, territorial governance and territorial development are interconnected concepts, and this is very important; effective territorial governance should have a positive impact on territorial development through a variety of aspects such as equitable economic development, urban planning, social welfare, building sustainable and inclusive cities, towns and rural localities, environmental sustainability, and the management and delivery of services.

The European Union serves as a successful model for good governance to development through its emphasis on a territorial approach that tailors policies to the unique needs and specificities of different regions. The European Union has developed the territorial agenda for 2030 (European Union 2020) in which the need to adopt policies built on the territorial dimension is emphasized. With this in mind, the EU states are urging their internal entities to adopt "policy responses with a strong territorial dimension and coordinated approaches acknowledging and using the diversity and specificities of places." (European Union 2020). Through new modes of governance and the involvement of local actors, policy-makers can tailor their strategies based on a spatial approach (Stead 2013: 142–144). Three dimensions are crucial in understanding the different approaches of territorial governance (European Parliament Committee 2015):

1. Stressing the territorial or place-based dimension of policymaking, i.e., reflecting the territorial specificities of the area concerned.
2. Bringing together players from different sectors and levels of governance.
3. Looking at the situation strategically and considering the long-term ramifications in order to achieve societal objectives.

Given the above, the territorialization of public policies necessitates continuous coordination and consultation at multiple levels, giving each territory a distinct configuration as an active and vital entity (Stead 2013: 143–144). The goal is to move from public policy to public action, adapting national orientations and sectoral policies to the specificities and potential of localities. "The underlying idea is that government institutions no longer have a monopoly on public action, which today involves a multiplicity of actors whose capacity for collective action determines the quality" (Duran 2001: 370). This is how territorial governance can be effective: by making the territory more viable, competitive, and equitable. In doing this, we will promote the "developmental role" of local governments (Schoburgh and Chakrabarti 2016) in implementing policies and managing resources while considering specificities of the territory. Therefore, the agendas of governance and local development intertwine due to the relationship of the two (Schmitt and Van Well 2016: 10–12), and a development-oriented government is required to ensure that socio-economic and spatial disparities are addressed, and specific developmental goals are achieved.

Nevertheless, it appears that one of the issues that international organizations have been addressing is how to tackle development in light of the UN's Sustainable Development Goals for 2030 (United Nations 2015), and how to adapt policies to the specific needs of territories. Indeed, the needs of individuals vary from one region to another depending on the availability and accessibility of resources which, in turn, are determined by the governance model and power dynamics within the territory.

1.2.3 Adaptability of Territorial Governance and Development Agendas in Africa and Middle East

It is worth noting that the territorial governance models vary broadly depending on the complexity of historical, cultural, and geopolitical contexts of each region. These models of governance also change over time as countries evolve and develop, and while some may have made significant progress, others may still be struggling. This is how territorial governance will open the way for action beyond simple planning. It is adaptable and inclusive, whereas previous governance models were built solely on economic or social factors. The power of this model lies in the fact that it is the go-to model used by states as it guarantees full control.

How governance relates to development across Africa and the Middle East is constrained by a variety of factors that have shaped, and continue to shape, the environment of states. In this volume, we will take an in-depth look at how aspects of territorial governance, i.e., participation and engagement, access to resources, meet local challenges, challenges often considered obstacles to development. We hope this demonstrates the inefficiency of adopting a single approach to fit the whole development agenda of entire world regions (Stead 2013). The case studies included here will allow the reader to apprehend the common issues, challenges, and strategies as follows:

- Decentralization and devolution: many countries have reformed public management and political decision-making to give local governments more powers and control over their territories.
- Rapid urbanization and rural poverty: many developing countries must deal with massive urban growth and increase in demand for essential services, while the urban-rural divide keeps widening. The major factor here is the lack of coordinated territorial governance strategies.
- Resource distribution: equitable distribution can be a real challenge for countries with significant regional disparities, limited resources and funds, and deficient infrastructure.
- Capacity building: local governments face significant challenges regarding the capabilities and competences of their personnel to manage structures, and to implement and follow-up projects and programs. The needs in terms of technical assistance, trainings, and practical knowledge are significant.
- Infrastructure development and service delivery: investing in adequate infrastructure and providing essential services such as water, sanitation, education, and healthcare is crucial for territorial and social development as well as economic growth, especially in remote and underprivileged areas. It also serves as a means to ensure spatial justice.
- Land tenure and property rights: disputes and conflicts over land and property rights are complex and hinders effective territorial governance.
- Lack of transparency and accountability: both of these can hinder effective administration and lead to mismanagement of resources.
- Participation and inclusion: participatory approaches can help address local needs and concerns by involving citizens, including women, refugees, and youth in decision-making and development planning.
- Cultural and ethnic diversity: territorial governance needs to be inclusive and accommodate the needs of people with different ethnic, linguistic, and cultural backgrounds to ensure its effectiveness.
- Conflict and instability: both have a significant negative impact on the territorial governance system and exacerbate vulnerabilities.
- International aid and cooperation: development cooperation can influence territorial governance models and systems as it benefits from foreign aid to strengthen projects and programs.

1.3 Structure of this Volume

This volume is divided into four parts. The first explores the institutional and legal frameworks of local governance which pertain across Africa and the Middle East generally, and illustrates their shortcomings through a number of case studies. In Chap. 2 Right to Development Approach to Local Governance in Africa, Carol Chi Ngang and Christopher Dick-Sagoe situate the right to development, as enshrined in the African Charter on Human and Peoples' Rights, at the center of the discussion. According to this Charter, national governments in Africa are under an obligation to ensure socio-economic and cultural development for all their peoples. While the charter has been universally adopted, governments have mostly failed to put in place the mechanisms—legal and administrative procedures, organizational entities able to ensure transparency, accountability and citizen participation—necessary in order for the right to development to become effective. In the absence of such enabling procedures, it is left to civil society organizations to advocate for local communities in the courts of law. The authors analyze two cases: Oku white honey production in Cameroon, and the Maasai cattle economy of Kenya and Tanzania, to illustrate how localized resources which could potentially foster equitable local development of rural communities are instead being developed in manners that exacerbate inequalities.

Robert Home makes a similar argument about urban communities in Chap. 3 Urban Governance and Climate Action Challenges in Africa. Using a human rights approach, Home relates how progress on achieving Sustainable Development Goals 11 (sustainable cities) and 13 (climate action) is impeded by dysfunctional laws on land and local administration. The author offers a list of the constitutional and legal reforms necessary if local governments are to fully contribute to achieving these SDGs. Some technical solutions to better land governance would include better urban planning, vulnerability assessments, and improved water, sanitation, and infrastructure. Institutional reforms would require more robust co-operatives and community land development trusts. Traditional community and land management authorities, which have proved more resilient in the political landscape than national technocratic ones, are also better at ensuring community participation in development efforts.

In Chap. 4 Regional Development and Territorial Competiveness, El Hassan Farhat and Khadija Darmame use the case of Morocco to highlight the potential of scientific research, and of technopoles in particular, in fostering inclusive regional development. Current Moroccan policies of advanced decentralization offer the opportunity to institutionally couple the output of research institutes and universities, and particularly of high-tech industries and R&D, to the needs of local economic actors.

In the next chapter, André Akono Olinga offers a comparative analysis of four francophone African countries: Cameroon, Gabon, Burkina Faso and Côte d'Ivoire, to argue for the importance of financial transparency in local governance if sustainable development is to be successfully pursued. Since 2010, all four states have enacted national laws to enhance transparency in public finance. These laws explicitly link transparent public finances to the consolidation of local democracy and local sustainable development, and all operate under conditions of decentralization of government services. According to Akono Olinga's findings, as local authorities have acquired increased jurisdictions, they have become increasingly politicized. The frailty of the institutional frameworks meant to control the management of local public finance results in a lack of transparency. There thus remains a need to strengthen the normative legal framework of sustainable development. The author advocates for the creation of monitoring systems for local public policies. On the one hand, there should be dynamic, flexible controls of local public finance, ones that take account of local social and geographical realities and that can be adapted over time. Mostly, there needs to be political control of local public finances through citizens' control of local governance, i.e.: through improved participatory democracy.

Mohamed El Mensi provides a longitudinal analysis of the institutional development of local authorities in Tunisia in Chap. 6. A centralized unitary state since independence in 1956, Tunisia's constitutions and laws have provided for various degrees of administrative decentralization. Because of the inherently authoritarian nature of public administration, which has prioritized security and control, under-resources local authorities have struggled to fulfill their role as agents of local development. The 2014 Constitution, fruit of the 2011 change of regime, set the terms for more effective local autonomy, yet it maintained the ultimate subordination of local governments to interference by the central State's administrative apparatus. Whatever positive outcomes could have emerged from this new legal framework, however, were further forestalled by a return to more overt structures of central control spelled out in the 2022 Constitution.

The first part of the volume concludes with an assessment of post-Arab Spring environmental activism by Zeina Moneer. In the wake of the Arab Uprisings, numerous environmental movements flared up across the region reflecting changing opportunities for activists as well as peoples' perceptions of flagrant socio-economic and environmental injustices. The conditions that influenced the praxis of environmental activism are explored, with a particular focus on the development of three major environmental movements: Algeria's 'No to Fracking' movement, Lebanon's 'You Stink' movement, and Morocco's 'We Are Not Trash' movement. These environmental movements were not a mere cry for a healthy environment. They underpinned deeper thinking about how nature and humans are interlinked in the everyday struggles to secure environmental, political and economic rights.

Part II of this volume delves into the practices of local governance as revealed by three specific instances of urban planning in large metropolitan areas. In Chap. 8, Fuad Malkawi demonstrates how the supposedly a-political practice of spatial planning of the Greater Amman Metropolitan Area was in fact the instrument of a political objective, that of transferring control of the Jordanian metropolis from elect local authorities and investing it unelected planners. Normative planning practices, modeled on those of post-WWII Britain, were presented as technical solutions to a range of urban development problems, effectively creating a "good planning" ideology whereby "experts" could and did over-ride municipal bodies, and where the public was entirely excluded.

Khadija Darmame, Abdelkader Kaioua and Eric Ross present a similar analysis of Casablanca's development as the prime metropolitan area of Morocco in Chap. 9: The Endless Challenges of Local Governance in Casablanca. Over 50 years of administrative and territorial reform have failed to orient the growth of the metropolis towards equitable or inclusive growth. A "deficit of governance" lies at the heart of the urban development challenge. The various stakeholders are unable to carry out their responsibilities according to their respective missions. Power relationships and cross-purpose decision-making enmesh public administrative agencies. Their prerogatives overlap; actions are needlessly duplicated, and policies are feebly implemented and followed up at the local level. The deficit is compounded by the fact that the elected bodies most representative of popular will are the least empowered vis-à-vis the top-down modus operandi of State agencies. If Greater Casablanca is to cope with urban growth, and ensure effective local participation in the decision-making, there is an urgent need to look at new forms of urban governance which can enhance innovation and collaboration between all actors and sectors.

In Chap. 10, Pascal Rey and Margot Petitpierre demonstrate how international norms and standards on governance, particularly the Equator Principles related to environmental and social impacts of development projects, can effectively promote good local governance. They study four differently-financed urban infrastructure projects in Conakry, Republic of Guinea. They find that the extent to which a given project adheres to the Equator Principles depends mainly on the willingness of the funding institutions to implement them. International agencies such as the World Bank, which has the most political leverage, are able to impose such best governance practices as stakeholder participation and impact assessments. On the other hand, funding agencies that are only accountable to national or local governments, which can ignore or over-ride national legislation, are poor guarantors of citizen participation and equitable compensation for the loss and damages a project may cause.

Part III of this volume explores the issue of access to public urban goods, and to housing and public space in particular. In Chap. 11, Esther Danso-Wiredu compares access to housing, water and sanitation in four poor neighborhoods in Ghana. The author finds that, whereas the Ghanaian government is unable to provide these to low-income citizens, non-state actors, such as traditional courts and authorities and, especially, local voluntary associations, are meeting the challenge. Furthermore, these non-state actors have enabled local self-government in these communities. In particular, they rely on social capital and on local citizen and stakeholder participation. Yet, despite their success in preventing poor neighborhoods from sinking into chaos, the paucity of resources of these service providers means that they are not sustainable. The State now needs to follow suit by enacting the kinds of policies capable of improving the housing and community infrastructure in poor communities.

In Chap. 12, Safaa Monqid describes the changes and innovative strategies that urban women in the Middle East have adopted in order to reclaim and transform their cities. Two main areas are explored: women's access to rights and decisions in public life, and the methods they use to appropriate public spaces. Women have always played an essential role in family, community, local and national organization and their collective action has always been important even if it is not always formal. Despite the successful mobilization of women to access public space, or, more likely, because of it, city administrations across the region are reinforcing barriers. Here, we have a clear case of the official institutions of governance resisting popular pressure to conform to the accepted norms of "good" governance.

This volume concludes in Part IV with three studies related to local resource management. In Chap. 13, Ahmed Haj Asaad and Khadija Darmame describe how the mismanagement of ground water and natural springs, already pronounced prior to the outbreak of the Syrian conflict in 2011, has been weaponized since. Poor water management, exemplified by the Ras al-Ain springs in the Jazeera region, has not only exacerbated the crisis in accessing water but has also led to the collapse of the fragile inter-social relationships local populations had developed over decades to strengthen interconnection and coexistence. This came to a head in 2020, when the water supply from these springs was cut for hundreds of thousands of residents of Al Hassakah and environs. The authors argue that, instead of wielding access to water as a weapon in the on-going political and military struggle between opposing national parties, the focus should be on instrumentalizing it in pursuit of social cohesion, as a means of restoring shattered relations between social groups.

In Chap. 14, Karen Rignall, Atman Aoui and Moulay Ahmed el Amrani analyze the governance capacity of local authorities when it comes to two large-scale extractive projects: copper mining and solar energy generation. They find that, contrary to its stated intention, Morocco's 2015 mining code fails to break with colonial-era wealth extraction policies built on conquest and expropriation they designate as "extractivism". The full cycle of permitting for exploration, exploitation, compliance, and decommissioning reveals contradictions in jurisdiction that complicate territorial governance for communes trying to plan comprehensively for economic development. The relations between residents, local elected officials, the state, and companies are still skewed in favor of the latter. The 2015 law perpetuates opacity at every level of the extraction process, from the granting of permits, to the disclosure of revenue flow, to the requirement to mitigate impacts. Attempts at redress are met with repressive measures. Democratizing knowledge about the legal framework of extraction can offer some tools for local actors, but these need to be "translated" in order for people not versed in high-level policy-making to be able to use them.

In Chap. 15, Samira Idllalène scrutinizes another extractive industry, namely sand mining. Sand procurement, essential for the construction sector, is a lucrative activity for local entrepreneurs in coastal areas. Since the enactment of the Coastal Act in 2015 Morocco's coastal zones are subject to a "integrated coastal zone management" policy which, in principle, requires a participatory approach and the establishment of a two-tier (national and regional) planning system. So far, few of these provisions have been put in place. Coastal municipalities, especially rural ones where most sand mining occurs, lack of financial resources and legal empowerment necessary to manage their coastlines (the sea remains under the jurisdiction of the national government). Meanwhile, despite this law and other pieces of legislation, illegal sand-mining, that is unauthorized extraction, continues apace. One possible avenue to address the dysfunction, proposed by the national Economic, Social and Environmental Council, is to replicate the Marchica Lagoon Agency in Nador, on Morocco's Mediterranean coast, or the Bouregreg Agency in Rabat-Salé. Yet, these types of all-powerful ad hoc national agencies directly contravene Morocco's official policy orientations promoting decentralization, the empowerment of local government, stakeholder and citizen participation, and integrated management of territories and resources.

This volume on local governance and development in Africa and the Middle East concludes with an assessment by Robert Home of the most salient issues it has raised.

References

Aïdi H, Lynch M, Mampilly Z (2020) Introduction: a transregional approach to Africa and the Middle East. In: Africa and the Middle East: Beyond the Divides, POMEPS Studies 40. Retrieved from: https://pomeps.org/introduction-a-transregional-approach-to-africa-and-the-middle-east

Davoudi S, Evans N, Governa F, Santangelo M (2008) Territorial governance in the making: approaches, methodologies, practices. Boletín de la Asociación de Geógrafos Españoles 46: 33–52

Duran P (2001) Action publique, action politique. In: Leresche JP (ed) Gouvernance locale, coopération et légitimité. Pédone, Paris, pp 369–389

European Parliament Committee (2015) Territorial governance and cohesion policy. Policy Department of Structural and Cohesion Policies. Retrieved from https://www.europarl.europa.eu/RegData/etudes/STUD/2015/563382/IPOL_STU%282015%29563382_EN.pdf

European Union (2020) Territorial Agenda 2030: A future for all places. Informal meeting of EU Ministers for Spatial Planning and Territorial Development. Germany. Retrieved from https://territorialagenda.eu/wp-content/uploads/territorial_agenda_2020.pdf

Hooghe L, Marks G (2016) Community, scale and regional governance: a Postfunctionalist theory of governance, vol 2. Oxford University Press, Oxford

Houston DJ, Harding LH (2013) Public trust in government administrators. Public Integrity 16(1): 53–76

Leloup F, Moyart L, Pecqueur B (2005) La gouvernance territoriale comme nouveau mode de coordination territorial. Géogr Econ Soc 7:7–331

Middell M (2017) Are transregional studies the future of area studies? In: Mielke K, Hornidge A-K (eds) Area studies at the crossroads: knowledge production after the mobility turn. Palgrave Macmillan, New York, pp 289–307

Oosterhof PD (2018) Localizing the sustainable development goals to accelerate implementation of the 2030 agenda for sustainable development. Governance Briefs 33. Retrieved from https://doi.org/10.22617/BRF189612

Pasquier R, Simulin V, Weisbein J (eds) (2007) La gouvernance territorial: Pratiques, discours et théories, Paris, L.G.D.J., 235 p.

Schmitt P, Van Well L (2016) Territorial governance across Europe: pathways, practices and prospects, Oxon, UK: Routledge

Schoburgh ED, Chakrabarti B (2016) Developmental local government: from concept to praxis. In: Schoburgh ED, Martin J, Gatchair S (eds) Developmental local governance, International political economy series. Palgrave Macmillan, London, pp 24–50

Stead D (2013) Dimensions of territorial governance. Plan Theory Pract 14(1):142–147. https://doi.org/10.1080/14649357.2012.758494

Stead D (2014) The rise of territorial governance in European policy. Eur Plan Stud 22(7): 1368–1383. https://doi.org/10.1080/09654313.2013.786684

Torre A (2010) Jalons pour une analyse dynamique des Proximités. Revue d'Économie Régionale et Urbaine 3:409–437

Torre A (2011) Les processus de gouvernance territoriale: L'apport des proximités. Pour 209-210: 115–122

Torre A, Traversac J-B (eds) (2011) Territorial governance: local development, rural areas and agrofood systems. Springer, Heidelberg/New York

United Nation Development Program (2010) Evaluation of UNDP contribution to strengthening local governance. Retrieved from https://www.oecd.org/derec/undp/47871446.pdf

United Nations (2015). Transforming our World: The 2030 Agenda for Sustainable Development. 70(1). Retrieved from https://sustainabledevelopment.un.org/content/documents/21252030%20 Agenda%20for%20Sustainable%20Development%20web.pdf

United Nations Industrial Development Organization (2016) The Role of Technology and Innovation in Inclusive and Sustainable Industrial Development: Industrial Development Report. Retrieved from https://www.unido.org/sites/default/files/unido-publications/2023-03/IDR-201 6-REPORT-en.pdf

Part I
Institutional and Legal Framework of Local Governance

Chapter 2
Right to Development Approach to Local Governance in Africa

Carol Chi Ngang and Christopher Dick-Sagoe

2.1 Introduction

Africa is relatively underdeveloped, and underdevelopment especially, manifests on a broad scale in local communities. In this chapter, we make the argument for a right-to-development approach to local governance in Africa for two compelling reasons. On the one hand, that local governance provides an appropriate framework for the realisation of the right to development and, on the other hand, that the right to development can, in turn, significantly enhance local governance. The proposition is motivated by the fact that African state governments are under a legal obligation within the framework of the African Charter on Human and Peoples' Rights, to ensure that socio-economic and cultural development is guaranteed to all the peoples of Africa. Among other measures, the obligation embodies a directive responsibility for policy making, which entails strategic rethinking on how to structure the mechanisms of governance so that suitable policies for redressing complex and compounding impediments to development at the local level can be crafted and implemented. Because they so adversely impact the human right to development, these impediments are often the subject of national-scale development plans.

The right to development is guaranteed to all the peoples of Africa. They are collectively entitled to socio-economic and cultural development with due regard to

C. C. Ngang (✉)
National University of Lesotho, Roma, Lesotho

Research Fellow, Free State Centre for Human Rights, University of the Free State, Bloemfontein, South Africa
e-mail: cc.ngang@nul.ls

C. Dick-Sagoe
University of Botswana, Gaborone, Botswana
e-mail: sagoecd@ub.ac.bw

© The Author(s), under exclusive license to Springer Nature Switzerland AG 2024
K. Darmame, E. Ross (eds.), *Local Governance and Development in Africa and the Middle East*, Local and Urban Governance,
https://doi.org/10.1007/978-3-031-60657-1_2

the equal enjoyment of their common heritage (African Charter 1981: Art 22(1)). The UN Declaration on the Right to Development (1986: Art 1(1)) adds a political dimension to its definition. If development is to be achieve as a human right, political considerations ought to be given as much attention as socio-economic and cultural ones. It is worth highlighting that issues related to human rights and development (which constitute the crux of the right to development) are primarily the responsibility of the state. Accordingly, the right to development requires national governments to take concrete measures to put in place mechanisms for its realisation, and eliminate impediments to development (DRTD 1986: arts 3(1) & 3(3)). This obligation mandates African governments to shape the national framework for development appropriately and in the process, ensure that human rights are respected, protected and fulfilled. This is seldom the case in Africa where implementation of the right to development is still largely aspirational and elusive. This is partly because of the lack of functional governance mechanisms capable of producing satisfactory development deliverables, especially at the local community level.

Local governance is defined here according to Anwar Shah and Sana Shah (2006:1), as the "formulation and execution of collective action at the local level" which, as the World Bank (1992: 3) clarifies, allows for the effective management of available resources for purposes of creating development. The idea of local governance, accordingly, aligns with the concept of the right to development as it is formulated and enshrined in the African Charter principally as a collective entitlement, the exercise and enjoyment of which is guaranteed to African peoples. Furthermore, the peoples of Africa are collectively entitled to use the continent's wealth of natural resources (common heritage) to generate and sustain socio-economic and cultural development, and to share equitably the gains resulting therefrom. The right-to-development approach to local governance is thus, proffered as a workable model that envisages applying rights-based standards to development practice at the local level. The analysis is constructed on the basis that local governance and the right to development exhibit analogous development characteristics and indicators.

As a starting point, it is worth reiterating that the challenges to development in Africa, which include the abuse and violation of human rights, extreme levels of poverty, and the devastating effects of climate change among many others, are most prevalent at the level of local communities. We posit in this regard that the right to development approach to local governance, if adopted and applied effectively, has the potential to equalise opportunities for development and accordingly, multiply prospects for sustaining collective initiatives for development to the benefit of local communities. In corroboration of this conviction, the section that follows, explores the local governance framework in determining what it envisages to accomplish, and in accordance, provides justification in support of the right to development as a suitable approach for the attainment of that purpose. The discussion, thereafter, illustrates how, with respect to taking concrete action, strategic localisation of community-based initiatives could promote and facilitate implementation of the right to development and in turn, activate effective local governance across Africa.

2.2   Understanding Governance and the Purpose of Local Governance

In this section, we explore the concept of local governance with the aim of eluci-
dating what it sets out to achieve, principally to support development at the local
level. The analysis is then juxtaposed with the normative concept of the right to
development as enshrined in the African Charter, which all 54 member states of the
African Union have ratified with the exception of Morocco. In accordance, we
reiterate the argument that in order to be effective, local governance ought to be
infused with the idea of the right to development, which is due and claimable by all
the peoples of Africa, including those in local communities, as a collective
entitlement.

Governance is a complex, elastic and versatile concept whose definitions vary
according to context (Vymětal 2007). For the purpose of this chapter, governance is
conceived within the larger context of governmentality, which has to do with the
way people are governed. In contrast to disciplinarian (coercive) forms of exerting
governmental authority, governmentality is concerned with improving the welfare
conditions of the population and hence, necessitates their active involvement and
unfettered participation in the processes of governance (Li 2007: 275; Foucault
1991: 100; Huff n.d.). In this way, governance has the potential, on the one hand,
to expand the choices available to people, including especially poor and vulnerable
groups that are at the centre of development and, on the other hand, to protect the
opportunities future generations are entitled to (UNDP 1997). However, because
citizens, groups, and organisations inevitably have difference interests, as Ledivina
Cariño (2006) draws attention to, it is noted in the UNDP Human Development
Report (1997) that governance aims to facilitate the mediation of those differences in
pursuit of the collective good. Thus, inherent in the concept of governance are norms
and processes that provide the opportunity to assert rights and obligations, promote
interests, and fulfil obligations.

Conceptually, governance is understood to be a much broader notion than
government and is defined as a process that equips elements in society with the
power and the authority to enact policies and influence decision-making concerning
public life and socio-economic development (Bonfiglioli 2003: 18). From a functional point of view, governance denotes a set of mechanisms, procedures and
organisational entities that are intended to facilitate and coordinate interactions
within society and to manage public resources for the common good. It is anchored
in the multifaceted processes of decision-making, which ought to include transparency, inclusivity, participation, accountability, respect for the rule of law and
responsiveness to inherent entitlements, legitimate expectations and the overall
exigencies of society. It involves the application of governmental norms and policies
within a system of constructed values and models that are leveraged by the dynamics
of power relations between governments and the governed.

The World Bank (1992: 3) provides a more lucid definition of governance as
embodying "the manner in which power is exercised in the management of a

country's economic and social resources for development." Lynda Bourne (2014: 1) conceives of governance as a framework for creating balance between economic and social goals, necessitating stewardship, accountability and the efficient use of resources with the aim to align, as far as possible, competing interests among stakeholders. It entails the exercise of governmental power in the management of public affairs. Peoples are justified not only to aspire to, but to demand a governance system that protects their rights and responds to their needs and interests. Governments, for their part, are obligated to be accountable to citizens and to promote their participation (USAID n.d.). Due to asymmetries between governments and the people they govern, this is hardly the case in Africa, where there generally is a dearth of local governance mechanisms. As a result, it is left to civil society organisations to advocate for local communities in the courts of law like in the SERAC case against the government of Nigeria (2001) where the Ogoni community was represented by the Social and Economic Rights Action Centre and the Centre for Economic and Social Rights – USA, the Endorois case against the government of Kenya (2009) where the Endorois community was represented by the Centre for Minority Rights Development and Minority Rights Group International, as well as the Ogiek Community case against the government of Kenya (2017) where the Ogiek people were represented by the Ogiek People's Development Programme and Minority Rights Group International.

To understand governance, therefore, also entails the interaction between government and civil society (organised society outside of the public sector) in shaping the functionality of government regarding the implementation of socio-economic and cultural development programmes and public sector investment (Wilson 2000). For governance to be effective, formal and informal relations ought to be established between government and civil society organisations. Furthermore, in order to influence governmental action with respect to ensuring transparency, accountability, citizen participation and the protection of human rights, civil society organisations must be sufficiently robust and have the capacity to advance societal priorities. It is at the level of local communities that governmental authority is exercised in closest proximity to citizens. It impacts the livelihoods of both individual citizens and entire communities.

The devolution of governmental powers to local communities is a prerequisite if local governance is to become practicable and effective. This requires profound decentralisation reforms. Wilson (2000) rightly notes that in many countries around the world, decentralisation results in the attribution of responsibilities to local governments, which tends to deepen interaction between the citizenry and government. Local government and the practice of local governance, as Aurora Ndreu (2016: 5) explains, enables people within a local jurisdiction to act collectively, with sufficiently close interaction and a common purpose to achieve desired welfare outcomes. Focus has increasingly shifted from centralised forms of operation to greater emphasis on local governance with a people-centred approach that allows for more efficient management of resources; notably, the need to empower local communities and to eradicate poverty, which is particularly important when vulnerable and economically disadvantaged groups like indigenous peoples and rural women are involved (Bonfiglioli 2003: 17–20; Bossyut and Gould 2000: 1–8).

In practical terms, local governance means the managing of governmental processes (exerting influence and control, administering rules, regulating interactions, providing direction) to benefit citizenry at the scale of localities. Local governance, therefore, also means devolution of governmental authority, and the mechanism through which that authority is exercised, to local communities so that local populations can constructively participate in making contextually suitable choices. For any human development to be sustainable, it is imperative that local people be at its centre. When development is conceived at the grassroots and supported by use of available resources, it has the potential to secure community ownership. It is more likely to be sustained than when it is conceived at the upper echelons of the governmental hierarchy and handed down from there. Local governance is a core aspect of political development, which is a component of the right to development. Thus, for local governance to be effective, it ought to be infused with the idea of the right to development.

2.3   Rights-Based Entitlement to Development

In addition to their other functions, African national governments are obligated to create the enabling environment for development to take place. Development is referred to in the broadest conceivable terms (including its infrastructural, technological, political, economic, socio-cultural dimensions). Ultimately this should translate into better standards of living and improved conditions of life for all. Emmanuel Ojo (2016: 92) observes in this regard that development must aim to achieve "qualitative and quantitative improvement in people's living standards [...]." With respect to this refined perception, the United Nations Development Programme (UNDP) maiden human development index report, provided a comprehensive definition of the concept of development as having a people-centred focus. Development is a process that should enhance people's productive capabilities and broaden their livelihood choices. This UNDP Report notes that "the process of development should at least create a conducive environment for people, individually and collectively, to develop their full potential and to have a reasonable chance of leading productive and creative lives in accord with their needs and interests" (UNDP Report 1990: 1).

Subsequently, the nebulous concept of development has evolved into a human right to which each and every human being and all people are legitimately entitled. Despite perceptions to the contrary, development is now universally recognised as a human right, formulated in concrete terms as the right to development (Marks 2013: 23, 26; UN Human Rights 2023: 495; Sengupta 2004: 180–183; M'baye 1972: 505). However, there remains some uncertainty about the conceptual clarity and legal nature of such a right. Being a human right means that development is central to well-being. It ought to be pursued relentlessly to achieve the full realisation of ancillary rights and freedoms. The Declaration on the Right to Development (DRTD) (1986: art 1) affirms that the right to development is an inalienable human right that entitles very human being and all people to participate in, contribute to, and enjoy economic, social, cultural and political development. The right to development implies the right of peoples to self-determination, and to sovereign ownership over their natural wealth and resources. The Declaration adds that "[t]he human being is the central subject of development and should be the active participant and beneficiary of the right to development' (DRTD, 1986: art 2(1)).

The right to development is acknowledged as an integral part of the corpus of universally recognised human rights. In 1993, the World Conference on Human Rights reached a consensus and adopted the Vienna Declaration and Programme of Action. Being an inalienable human right means that ensuring the right to development, as with every other human right, is "primarily the responsibility of Governments" (United Nations World Conference on Human Rights 1993: para I(1)). The DRTD (1986: art 3(1)) states that "[s]tates have the primary responsibility for the creation of national and international conditions favourable to the realization of the right to development." To achieve this purpose, national governments have the duty to adopt national development policies. It is stipulated that:

> States have the right and the duty to formulate appropriate national development policies that aim at the constant improvement of the well-being of the entire population and of all individuals, based on their active, free and meaningful participation in development and in the fair distribution of the benefits resulting therefrom (DRTR (1986: art 2(3)).

Legal provisions are generally not self-executory; they need mechanisms through which effective implementation can occur. The policy obligation that accrues from the above provision implicitly enjoins national governments in Africa not just to adopt requisite policies to achieve development, but also to put in place the kinds of governance mechanisms that can support and facilitate such policies. There is an imperative need for governance mechanisms at the level of local communities. Mechanisms of local governance are imperative so that "no one is left behind" in the development process, as stated in the UN Sustainable Development Goals (SDG) (2015: para 4).

While the DRTD and the SDG are in principle non-binding, and therefore impose no legally enforceable obligation on governments, the African Charter imposes a legally binding duty on state parties to implement the right to development for all the peoples of Africa. Accordingly, having ratified the African Charter, African governments are compelled to ensure that the appropriate context is established for the right to development to be exercised. To this end, Article 22 of the Charter provides for the following:

1. All peoples shall have the right to their economic, social and cultural development with due regard to their freedom and identity and in the equal enjoyment of the common heritage of mankind.
2. States shall have the duty, individually or collectively, to ensure the exercise of the right to development.

As highlighted in the SDG framework document, the 'all peoples' specification in the African Charter affirms that as a matter of legal entitlement, essentially none of the peoples of Africa will be left behind. All should enjoy equitable access to opportunities for development, actively participating in and contributing to the development processes, and share in the benefits that obtain therefrom. This guarantee is of particular importance and worth reiterating because across Africa, local communities have often been left behind as far as development is concerned. It is at the level of local communities that socio-economic and cultural development have remained elusive, where impoverishment seems to be most entrenched, and where, more often than not, the effects of adversities are most strongly felt.

Article 22(2) of the African Charter stipulates the duty of state parties, either individually or in cooperation with others, to ensure that all the peoples of Africa have the opportunity to exercise the right to development. As noted earlier, it entails that national governments adopt the requisite national development policies and put in place the requisite governance mechanisms to ensure that the right to development can be exercised. The national development policy imperative implies that appropriate development policies must be put in place and must apply evenly across the national territory. This is problematic in the sense that the socio-economic and environmental contexts are usually not homogenous at the national level. It follows that a policy of general application will certainly not respond to and satisfactorily redress context-specific issues everywhere in the same manner. Thus, with the understanding that development exigencies are generally not uniform and that communities are not homogeneous, national development policies cannot be implemented uniformly everywhere and produce the same development outcomes in every locality.

For the populations in local communities not to be left behind in the development process requires that policymaking power is sufficiently decentralised to enable effective responses to issues that directly impact livelihoods. Policymaking is integral to governance, including at the local level. In addition to national development policies, local communities should have the opportunity and, indeed, be empowered to adopt context-specific local development policies that respond in a targeted manner to the realities on the ground. In this way, local governance is particularly focused on ensuring that development is treated as a human right—a right owed to often-neglected local communities—and consequently, that in responding to their specific development needs, basic human rights are respected. It is in this sense that our argument on the right-to-development approach to local governance is formulated.

Arjun Sengupta (2004: 180) defines the right to development as the right to a particular process of development wherein all fundamental rights and freedoms can fully be realised in a manner that is consistent with human rights standards. This means that the local development processes, as with other spheres of society, must accord with and adhere to right-to-development standards. In the African case, where the right is explicitly provided for in the African Charter, right-to-development standards require that the mechanism through which development is framed be conceptualised with the purpose to achieve the intended human rights and development objectives. Accordingly, the political dimension of the right-to-development requires a governance system that promotes the rule of law, transparency and accountability, and allows for genuine participation in shaping the development processes. The system should ensure (re)distributive justice in order that development gains can be shared equitably and with adequate responsiveness to the collective development aspirations of the entire community.

In practical terms, A local governance system implies that even as local communities are confronted with a myriad of development challenges, cognisance ought to be taken of the local socio-economic and cultural realities of communities. The experience of development varies greatly and, therefore, development cannot proceed in the "flat" mould of national policy solutions. Basing on this reasoning, we contend that even as the right to development is broadly guaranteed to all the peoples of Africa, and that state parties are enjoined to ensure adequate implementation thereof, generic national development policies may not be enough to provide solutions to the actual development problems and associated needs in every locality. Local communities generally have resources and capabilities at their disposal, which they can use and/or transform for their collective socio-economic and cultural development. Nonetheless, local communities remain impoverished. We contend that what holds them back is the absence of objective-driven local governance policy contexts for enabling the local communities exercise the right to development in a productive manner and for gainful purposes.

2.4 Strategic Localisation of Development Initiatives

As highlighted earlier, the governments of African states are under an obligation to formulate appropriate national development policies that aim at the constant improvement of the well-being of the entire population. To achieve this purpose, requires effective devolution of governmental authority to local communities so that they may, without constraints, participate in the processes for development and in the fair distribution of the benefits that result from them. It requires enabling local government authorities to formulate policies that are informed by local realities. In turn, such strategic localisation means equipping localities with a functional governance mechanism that allows for collective exploration of practicable possibilities in terms of available resources and the capabilities to initiate, drive and sustain development on a local scale.

Cognisance is taken of the fact that development is cumbersome, necessitating the mobilisation of enormous human, material and financial resources. While the theoretical argument in favour of a right-to-development approach to local governance is plausible as we posit, the central question is whether local communities in Africa have the resource capacity to self-sustainably create development that guarantees equitable benefits to the communities concerned. The analysis here draws from the concept of the right to development in Africa, which implores all peoples across the continent to look to their common heritage as an equitable means to achieve socio-economic and cultural development. In the subsection that follows, we provide two practical illustrations of local community initiatives with the potential to scale-up the socio-economic and cultural development benefits, improved well-being and better standards of living.

2.5 Some Practical Illustrations of Strategic Localisation

2.5.1 Case 1. Bee Farming in the Oku Community, Cameroon

The Oku community, which is host to the Kilum-Ijim forest highlands (Oku Mountain range) in the Northwest region of Cameroon, is known for bee farming and famous for the production of a rare type of honey that has been trademarked as "Oku White Honey" (World Intellectual Property Organisation (WIPO) 2015; Musiza 2021). Oku White Honey is a certified Geographical Indication Product (GIP). The GIP certification of the Oku White Honey is based on the environmental uniqueness of the region where it is produced, its unique quality and originality, and the traditional methods of production or the cultural and indigenous characteristics linked to its production (Konsum 2017). Although the Oku White Honey is produced in a remote part of Africa, its global market value is likened to other renowned GIP brands like Mexican tequila, Italian parmigiana, Colombian coffee (WIPO 2015), or Bordeaux wine (Konsum 2017). If bee farming and the production of the premium Oku White Honey is to be scaled-up to supply the global market on a scale comparable to these other GIP brands, it is likely that living standards for the local community will significantly improve. As Konsum notes, this potential remains untapped. The Oku community has a total population of about 130,000 people, and the honey's production belt is over 20,000 hectares (200 square kilometres). Bee farming is a major source of income generation and sustainable livelihood for the people in the community. The GIP certification and subsequent increased commercialisation of the Oku White Honey is reported to have increased incomes and brought significant positive economic and social development mostly, to organised bee farming groups involved in the production. As these organisations have few members, the wider community remains largely impoverished. It is assumed here that the trademarking and GIP certification of the Oku White Honey entitles the entire community to benefit from the activity, not just specific groups or individuals.

Article 22 of the African Charter guarantees the right of all peoples to socio-economic and cultural development with due regard to their freedom and identity and the equal enjoyment of the common heritage. Therefore, it is incontestable that everyone who identifies as a member of the Oku community is entitled to freely share in the benefits that obtain from bee farming, which is a common heritage of that community. Traditional knowledge of bee farming and production of the Oku

White Honey and related processes in the value chain, including the making of beehives, are acknowledged to have been passed down from generation to generation (WIPO 2015). The Oku community constitutes a people in accordance with the notion of peoples contained in article 22 of the Charter, which imposes an obligation on state parties to create enabling environments for them to exercise that right. The obligation necessitates the devolution of state authority so that the people in that locality are able to meaningfully participate in the honey production and share the benefits generated from the development process.

The fact that bee farming remains rudimentary and not organised in a manner that collectively benefits the entire Oku community may be attributed to the lack of an enabling local governance mechanism. A number of governmental, non-governmental and community-based entities among other stakeholders, are actively involved in organising and capacitating the local community in sustainable bee farming and honey production. By 2017, just over 1000 community members had received training in standardising quality honey production (Konsum 2017). We contend that at this rate, the efforts are relatively insignificant. The prospects for valorisation and expansion of the bee farming value chain remain huge and unexploited because of the absence of a suitable local governance mechanism to provide the Oku community appropriate direction. We are of the view that the lack of interest in bee farming shown by the greater part of the Oku community probably stems from lack of awareness of their entitlement to the right to development, which entails their involvement and participation in the endeavours and processes for development within the community (in this instance, bee farming as a common heritage) and in sharing equitably in the proceeds generated.

In addition to being bound under article 22 of the African Charter by virtue of its ratification in 1989, it is worth noting that the Cameroon Constitution adopted in 1972 (amended since) was the very first legal instrument to give statutory recognition and protection of the idea of the right to development (Ngang and Kamga 2018: 184 & 194). The Constitution states: "We, the people of Cameroon [. . .] resolve to harness our natural resources in order to ensure the well-being of every citizen without discrimination, by raising living standards, proclaim our right to development as well as our determination to devote all our efforts to that end [. . .]." The determination to devote all efforts to raising living standards and ensuring the well-being of the people of Cameroon, in fulfilment of the proclamation of the right to development, entails extending that commitment to local communities. It is the duty of government to create the enabling environment for the right to development to be actualised. For the Oku community, that requires a local governance mechanism that will adopt the requisite local development policies.

The local development policy framework ought to ensure that the Oku community is sufficiently informed and educated on the entitlements to bee farming as a common heritage. Each of its members is enjoined to participate in the production of Oku White Honey as a matter of right, for the collective benefit and shared prosperity of the entire community. Adequate local governance would ensure the development of the Oku community as a bee farming and honey production hub. This would not only expand but also equalise opportunities for socio-economic and cultural development within the community, in accordance with current understandings of development as a people-centred process of improving living conditions and raising well-being. The right-to-development approach to local governance guarantees that the Oku community can, and ought to be mobilised to productively engage in communal bee farming at different stages within the value chain, including in the manufacturing of beehives and related accessories, bee farming proper, honey production, branding and packaging, quality assurance and standardisation, and in the marketing and distribution processes.

2.5.2 Case 2. Cattle Economy of the Maasai Community, Kenya and Tanzania

The region that straddles the Great Rift Valley of southern Kenya and northern Tanzania has been inhabited from time immemorial by the Maasai—a predominantly pastoralist community—who have been the subject of historical marginalisation and victims of various human rights violations. The Maasai have not had their human rights claims adjudicated by any court or quasi-judicial enforcement mechanism, as has been the case with the Endorois and the Ogiek. However, like other indigenous communities in the east African region, the Maasai have reportedly been dispossessed, forcibly evicted and displaced from their ancestral lands (Human Rights Watch 2023; Minority Rights Group International 2023; International Work Group on Indigenous Affairs 2022; Galaty 2010). Thus, they have been systematically denied or deprived of many of their human rights entitlements, including the right to development guaranteed to them under article 22 of the African Charter.

With a total population of approximate 1.2 million, the Maasai are known to possess enormous wealth in the form of cattle, which they have religiously preserved as a communal heritage and as an integral part of their collective identity. National Geographic Magazine (2022) hails the Maasai for having established a flourishing cattle economy that thrives on barter-type transactions, with cattle also used as currency. However, despite efforts to protect their unique indigenous cultural heritage, in recent years the rapidly expanding commercial market economy has threatened the Maasai way of life. This has meant that their "[. . .] highly developed and ritualized barter system, organized around the currency of cattle, has had to give way to state legal and policy preferences in favour of a money-driven economy that is founded on nonindigenous concepts of property and value" (National Geographic 2022).

It is worth emphasising here that neither the governments of Kenya nor Tanzania can legitimately define what ought to constitute development for the Maasai. According to the African Charter, the right to development is to be achieved with due regard to the defining components of "freedom and identity." It makes clear that development can be determined only by the collective of the peoples concerned in accordance with the realities in their context (Ngang 2022: 57). The governments of Kenya and Tanzania have repeatedly encroached on Maasai lands for purposes of tourism, claiming to "conserve" them. While these policies may improve national-scale macroeconomic indicators, they have little to zero development potential for the Maasai. On the contrary, the findings of a 2015 livelihood survey indicate that "the main determinant of Maasai wellbeing can only be equated to livestock wealth" (Tiampati 2015: 4). If socio-economic and cultural development is to be achieved for the Maasai as guaranteed in the Africa Charter, the Maasai need to be allowed the freedom to use the resources at their disposal in making their own development choices.

As earlier noted, the African Charter guarantees that the right to development is achievable with due regard to the "equal enjoyment of the common heritage." For the Maasai community, cattle constitute a common heritage. Cattle are valued as symbols of wealth and social status as well as source of livelihood and sustenance. Cattle raising, therefore, ought to be enabled and allowed to develop for greater socio-economic and cultural gains of the community. The Maasai cattle economy is a pillar of economic development–accounting for at least 10% and 14% of the gross domestic product (GDP) of Kenya and Tanzania respectively (Tiampati 2015: 5). Yet, the cattle economy has not been of significant benefit in raising living standards and in transforming livelihoods for the Maasai. This is probably because national governments are more inclined to grab Massai land for "conservation" in order to boost the tourism industry.

One of the central tenets of the right to development is that it envisages equality of opportunity for development, which entails the liberty to choose between alterna-tives (Ngang 2022: 72). The right to development equally envisages equity and justice in development (Ngang 2022: 76), to the effect that conservation and tourism cannot reasonably be imposed on the Maasai when their preference is cattle raising. It is unjust and inequitable to do so and, in effect, it contravenes the right to development. With respect to the liberty to choose, the cattle economy provides better prospects to the Maasai than the tourism industry that is being forced on them. The advent of the African Continental Free Trade Area (AfCFTA) for instance, presents a huge opportunity (with a market size of an estimated 1.2 billion people) for the Maasai to explore in terms of scaling-up the cattle economy, which is shrinking under the pressure of the Kenya and Tanzanian government conservation policies. With sufficient support, the Maasai cattle economy has the potential to grow exponentially and become the principal supplier of cattle and dairy products across the African free market.

The Maasai have consistently claimed sovereignty and autonomy as a people (Meitemei 2010). The concept of "peoples" in Africa has been defined in academic literature (Kiwanuka 1988: 101) and in jurisprudence (*Gumne*, 2009: paras 169–179; *Endorois*, 2009: paras 146-157yy; *Ogiek Community*, 2017: para 208). Being a people qualifies and entitles the Maasai to a legitimate claim on the right to development. The yearning of the Maasai for sovereignty and autonomy literally implies an aspiration for self-determination (self-governance), which is ancillary to the right to development, and which is attainable through local governance. By law, the governments of Kenya and Tanzania (state parties to the African Charter) have the duty to create the enabling conditions, including decentralising the centres of power, that would allow the Maasai to take charge of governing themselves in their own localities. This is how they can exercise the right to development as envisaged in article 22(2) of the African Charter.

According to Legaspi (2005), local governance is anchored on the authority to formulate a regulatory framework that supports free, active and meaningful participation of various stakeholders in decision-making regarding governance of community resources and in the full development of the potential of community members, including defending their fundamental rights and freedoms and empowering them with the capabilities to shape their own destiny. If set in place, local governance would grant the Maasai community the decision-making powers needed to make informed development choices concerning how to manage and grow the cattle economy in a rights-based manner that guarantees equal enjoyment of their common heritage in cattle raising, which they have preserved from time immemorial, for the collective socio-economic and cultural development benefit and uplifting of the entire community.

2.6 Conclusion

This chapter has aimed to illustrate that local governance and the right to development are mutually reinforcing. On this basis, we have endeavoured to construct the theoretical argument for a right-to-development approach to local governance in Africa. Development is an all-encompassing process, rendered all the more complex to the extent that it is conceived as a claimable human right encompassing the socio-economic, cultural as well as political dimensions, which must be achieved in their totality. A legal obligation is imposed on African governments to deliver development to all the peoples of Africa as a human right. The development processes, and the way it is conceived and implemented, must comply with rights-based standards. As part of that obligation, national governments are obligated to put in place solid governance supported mechanisms to facilitate and enable the peoples of Africa, particularly those in local communities, to exercise the right to development guaranteed to them.

It is estimated that a satisfactory delivery of right-to-development expectations would have seen massive improvement in well-being and standards of living for the peoples of Africa. That has not been the case, despite the African origins of the idea of a human right to development and over 40 years of legal recognition and protection of such an entitlement in the African Charter. Implementation of the right to development requires the active involvement and meaningful participation of the peoples of Africa both in development decision-making and in the development processes proper. This compels national governments to ensure that the system of governance, particularly at local government levels is supportive of the people-centred, rights-based approach to creating development. A practical way to achieve this is through local governance. The fact that the promise of a right to development has not been achieved is attributed, for the most part, to the failure of African governments to fulfil their obligation to put in place contextually relevant local governance mechanisms to facilitate implementation of the right to development.

Across Africa, local communities exhibit strong unique characteristics, which they hold onto as communal heritage. These have the potential to transform standards of living within these communities and present opportunities to be explored for the collective enjoyment of the right to development as envisaged in the African Charter. The two practical illustrations discussed in this chapter are just examples of how community-based initiatives could be identified and strategically harnessed as localised development hubs. Cumulatively, local governance initiatives of this type would enable various localities across the continent to share equitably in the benefits that obtain from their common heritage. A major limiting factor has been the noticeable absence of functional local governance mechanisms to shape the direction for development within these communities, which has meant sustained underdevelopment.

The right-to-development approach to local governance, as suggested in this chapter, necessitates that African governments admit the many shortcomings of their undertakings under the African Charter and, in effect, take concrete measures to decentralise power and development decision-making so that local communities are equipped with the governance capacity to translate their local realities into development opportunities. For development to be sustainable, it must start at the local scale. The role of civil society organisations is crucial in drawing the attention of national governments to their obligations in creating the conditions for all peoples to exercise their right to development, which essentially, entails strengthening governance at the local level.

References

African Charter on Human and Peoples' Rights adopted in Nairobi, Kenya on 27 June 1981. OAU Doc CAB/LEG/67/3 Rev. 5 (1981)

African Commission on Human and Peoples' Rights v Republic of Kenya (2017) Appl No 006/2017 (Ogiek Community case)

Bonfiglioli A (2003) Empowering the poor: local governance for poverty eradication. United National Capital Development Fund, New York

Bossyut J, Gould J (2000) Decentralisation and poverty reduction: elaborating the linkages. In: Statute for development studies, policy management brief 12. University of Helsinki, pp 1–8

Bourne L (2014) The six functions of governance. PM World J 3(11):1–6

Cariño LV (2006) From traditional public administration to the governance tradition: research in NCPAG, 1952–2002. Philipp J Public Adm 50(1):1–22

Centre for Minority Rights Development (Kenya) and Minority Rights Group International on behalf of Endorois Welfare Council v Kenya Comm 276/2003 (2009) AHRLR 75 (ACHPR 2009) (Endorois case)

Declaration on the Right to Development, Resolution A/RES/41/128 adopted by the UN General Assembly on 4 December 1986

Foucault M (1991) Governmentality. In: Burchell G, Gordon C, Miller P (eds) The Foucault effect: studies in governmentality. University of Chicago Press, Chicago, pp 87–104

Galaty J (2010) Maasai land, law, and dispossession. Cultural Survival https://www.culturalsurvival.org/publications/cultural-survival-quarterly/maasai-land-law-and-disposses sion. Accessed on 14 June 2023

Huff R (n.d.) Governmentality. Encyclopaedia Britannica https://www.britannica.com/topic/governmentality. Accessed on 10 June 2023

Human Rights Watch (2023) Tanzania: Maasai forcibly displaced for game reserve. 27 April. https://www.hrw.org/news/2023/04/27/tanzania-maasai-forcibly-displaced-game-reserve. Accessed on 14 June 2023

International Work Group on Indigenous Affairs (2022) 70,000 Maasai in Loliondo, Tanzania, face another forceful eviction. https://www.iwgia.org/en/news/4597-maasai-loliondo-tanzania-force ful-eviction.html. Accessed on 14 June 2023

Kelvin Mgwangwa Gumne & Others v Cameroon Comm 266/2003 (2009) AHRLR 9 (ACHPR 2009)

Kiwanuka RN (1988) The meaning of 'people' in the African charter on human and peoples' rights. Am J Int Law 82(1):80–101

Konsum TK (transcribed by Glory Ogbeugbu) (2017) Climate change and livelihood: the case of Oku community, Cameroon. https://glowinitiative.org/climate-change-and-livelihood-the-case-of-oku-community-cameroon/. Accessed on 21 Apr 2023

Legaspi PE (2005) Overview of governance framework. A working draft on a Handbook on LGU-SPA Partnership; UP NCPAG

Li TM (2007) Governmentality. *Anthropologica* 49(2):275–281

M'baye K (1972) Le droit au développement comme un droit de l'homme: Leçon inaugural de la troisième session d'enseignement de l'Institut International des Droits de l'Homme. Revue des Droits de l'Homme 5(1):505–534

Marks SP (2013) The human rights framework for development: seven approaches. In: Sengupta A, Negi A, Basu M (eds) Reflections on the right to development. Sage Publications, New Delhi, pp 23–60

Meitemei O-D (2010) Maasai autonomy and sovereignty in Kenya and Tanzania. Cultural Survival https://www.culturalsurvival.org/publications/cultural-survival-quarterly/maasai-autonomy-and-sovereignty-kenya-and-tanzania. Accessed on 13 June 2023

Minority Rights Group International (2023) Beyond just conservation: a history of Maasai dispos-session. https://minorityrights.org/2023/02/23/beyond-just-conservation-a-history-of-maasai-dispossession/. Accessed on 12 June 2023

Musiza C (2021) A pathway to international registration for African geographical indications under the Geneva act: the case for Oku white honey. J Intellet Prop Law Pract 16(4–5):394–401

National Geographic (2022) The cattle economy of the Maasai. https://education.nationalgeographic.org/resource/cattle-economy-maasai/. Accessed on 21 Apr 2023

Ndreu A (2016) The definition and importance of local governance. Soc Nat Sci J 10(1):5–8

Ngang CC (2022) The right to development in Africa. Brill, Leiden/Boston

Ngang CC, Kamga SD (2018) 'O Cameroon, thou cradle of our fathers. . .: land of promise' and the right to development. In: Ngang CC, Kamga SD, Gumede V (eds) Perspectives on the right to development. Pretoria University Law Press, Pretoria, pp 182–202

Ojo EO (2016) Underdevelopment in Africa: theories and facts. J Soc Political Econ Stud 41(1): 89–103

Sengupta A (2004) Human right to development. Oxf Dev Stud 32(2):179–203

Shah A, Shah S (2006) The new vision of local governance and the evolving roles of local governments. In: Shah A (ed) Local governance in developing countries. The World Bank, Washington DC, pp 1–46

Social and Economic Rights Action Centre (SERAC) & Another v Nigeria Comm 155/96 (2001) AHRLR 60 (ACHPR 2001) (Ogoni case)

The World Bank (1992) Governance and development. The World Bank, Washington, DC

Tiampati M (2015) Maasai livelihood and household sources of revenue report. African Conservation Centre Survey Report

United Nations Department of Economic and Social Affairs (2015) Ensuring no one is left behind. https://www.un.org/en/desa/ensuring-no-one-left-behind-0. Accessed on May 2023

United Nations Development Programme (1990) Human development report 1990. Oxford University Press, New York/Oxford

United Nations Development Programme (1997) Reconceptualizing governance. Discussion paper 2. Bureau of Policy and Programme Support, management development and governance division, New York

United Nations World Conference on Human Rights (1993) Vienna declaration and Programme of action, UN Doc A/CONF157/24, 25 June 1993

USAID (n.d.) Promoting good governance. https://www.usaid.gov/democracy/promoting-good-governance. Accessed on 26 May 2023

Vymětal P (2007) Governance: defining the concept. Prague University of Economics and Business, Faculty of International Relations Working Papers, pp. 1–16

Wilson RH (2000) Understanding local governance: an international perspective. RAE - Revista de Administração de Empresas 40(2):51–63

World Intellectual Property Organisation (2015) Oku White Honey: Cameroon. https://www.wipo.int/ipadvantage/en/details.jsp?id=5554. Accessed on 7 May 2023

Chapter 3
Urban Governance and Climate Action Challenges in Africa

Robert Home

3.1 Introduction

The African continent faces great development challenges, which are complicated by its geographical size, diverse populations and languages, and by the legacies of past colonial rule. Its development prospects depend upon how its governments manage rapid population growth and the newly recognized existential threat from climate change. The problems are 'wicked' (in the sense of resisting solutions rather than evil), and their complexity means that efforts to solve one aspect may reveal or create other problems (Conklin 2001).

This chapter aims to provide an overview of the potential contribution of local government to the challenges of urban governance and climate action. It draws from an extensive literature on international development and planning, and particularly the growing body of work by African researchers, based both in the continent and the diaspora. The first section investigates issues of governance at different levels, international, national, local and community-based. The next explores the rise of human rights in the development agenda, which have been given new urgency by the Covid-19 pandemic and the emphasis upon 'leaving no-one behind'. The fourth section concerns over-arching issues of land in the second largest continent by land area: access to land and reforms of land law and planning. The next two sections focus on two of the current UN Sustainable Development Goals: SDG 11 on sustainable cities and SDG 13 on climate action. The conclusions draws the themes together and assesses prospects for the future.

This chapter is partially based on the contributions to the author's edited collection (2021c) *Land Issues for Urban Governance in Sub-Saharan Africa,* and his article (2021b) History and Prospects for African Land Governance: Institutions, Technology and 'Land Rights for All'.

R. Home (✉)
Department of Business and Law, Anglia Ruskin University, Cambridge, UK

K. Darmame, E. Ross (eds.), *Local Governance and Development in Africa and the Middle East*, Local and Urban Governance,
https://doi.org/10.1007/978-3-031-60657-1_3

3.2 Issues of Governance

The concept of 'governance' has emerged, as distinguished from 'government',
Political science had long concentrated upon the state as a sovereign authority that
embodies the general will of its people and is based upon certain fundamental
principles, but that approach has been challenged by a more radical and decentred
theory of the state (Bevir and Rhodes 2010). This sees the state as neither monolithic
nor a causal agent, but rather a product of diverse beliefs about public authority,
contending (and sometimes unstable) cultural traditions and practices, and contin-
gent actions of individuals. Governance scholarship is now moving towards interdisciplinary and post-disciplinary approaches to approaching problems. For
example, ethnographic and observational methods can cover beliefs and preferences,
while historical narrative methods can trace the development of traditions in
response to challenges. Such a frame of analysis brings people back into the study
of the state, and acknowledges diversity, contingency and ruptures in the traditions
and institutions by which authority is exercised, and the processes of interaction and
decision-making between those institutions. It recognises the importance of politics
and power while seeking to be conceptually neutral (Fukuyama 2013).

The network of governance institutions comprises not only the 'state' (central,
regional and local) but many other institutions, public and private, community-based
and traditional. The World Bank in 1996 initiated its Worldwide Governance
framework, identifying six dimensions to governance: voice and accountability,
political stability, government effectiveness, regulatory quality, rule of law, and
control of corruption (Kaufmann 1999). Good governance is seen as important for
achieving development goals, and its failings have serious negative consequences
for society as a whole.

At international level the UN has formulated various global development goals in
the twenty-first century, which grew from the eight 'Millennium Development
Goals' for the period 2000–2015 into seventeen Sustainable Development Goals
(SDGs) for 2015–2030 (Global 2030 Agenda 2015). The SDGs were accompanied
by many targets and monitoring indicators, and UN-Habitat identified the 2020s as
being 'the decade of progress' (UN-Habitat 2019c). This chapter concerns particularly SDGs 11 and 13, but others are also relevant, notably SDG 16 ('peace, justice
and strong institutions'), including growing 'civic space', and SDG 17 ('global
partnerships'). Coinciding with the Sustainable Development Agenda the African
Union (AU) adopted its own Agenda 2063, with a ten-year implementation plan that
included no less than seven aspirations, 20 goals, 13 fast-track projects, 39 priority
areas and 255 targets (AUC 2015). These embraced wide issues of democracy,
cultural identity and continental integration, and offered an ambitious and idealistic
vision of an African future of good governance, democracy, respect for human
rights, justice and the rule of law.

A helpful academic lens for exploring these issues is offered by the theory of
historical institutionalism, which investigates sequences of social, political and
economic behaviour and change over time, and their influence upon institutional

and political structures and outcomes (Mahoney and Thelen 2010). In the case of Africa, at independence countries kept not only their former colonial boundaries under the *uti possidetis* principle, but also a largely intact colonial legal and bureaucratic framework, and some states had legal systems that derived from more than one colonial power (Shaw 1996).

The SDGs are careful to respect each UN member country's policy space and leadership, and the primary responsibility for implementing them falls to sovereign nation states. The AU has 56 member states, the largest number of any continent, and their populations range from a few million to over two hundred million (Franzsen and McCluskey 2016). They largely maintain former colonial boundaries and often split ethnic groups across these boundaries. The European languages of former colonial masters remain official languages of administration in many African countries—English in twenty-two, French in twenty, and Portuguese in five. Hybrid languages (Swahili, pidgin, creole) allow different language-speakers to communicate with each other, but can allow Western epistemology and languages to continue largely unchallenged. Other official languages in Africa are Arabic (twelve countries), Amharic in Ethiopia, Kinyarwanda in Rwanda, and Somali in Somalia, while fifteen countries have more than one official language. Such linguistic diversity, comprising at least three hundred separate languages in five language groups, greatly complicates governance, and the anthropologists of the former colonial 'masters' treated African cultures as static, timeless and separate from the modern world (Hammond-Tooke 1997).

Sub-national level governance institutions can be either local/regional government or non-governmental/citizen-based organisations (NGOs/CBOs). They have been until recently neglected by both international and national institutions, yet it is 'on the ground' where such development challenges as urban resilience and climate action are experienced. In the colonial past local administration was typically undertaken by centrally appointed officials of the colonial power, and local authorities were weak and under-resourced (Mamdani 1996). The largest organization of sub-national governments in the world—United Cities and Local Governments—was founded in 2004 to lobby for an increased role and influence for them, through law reform and the decentralization of functions and funding principles (UCLG 2019; Garcia Pena 2023). In Africa fiscal relations between its growing urban areas and their local governments have often resulted in weak, fragmented and hybrid urban governance, and poor physical and social infrastructure. The UN's Addis Ababa Action Agenda 2015 proposed policy actions for member states to achieve the SDGs, but was weak on how fragmented systems for financing could be improved, and particularly the potential of land-based finance approaches (Cirolia 2021; UN-Habitat 2023). In Nigeria, Africa's most populous country, for example, the federal structure has resulted in a hyper-centralized federal government, partly rooted in colonial foundations, in which civil and military rule alternated in cycles, creating economic vulnerabilities, and multiple ethnicities that hinder decentralization (Rotimi 2022). To take another example, the constitutional architecture of the Republic of South Africa means that, while its people often experience food insecurity and malnutrition, local authorities have little statutory responsibility for such vital matters (De Visser 2021). In 2010 some 90% of Africa's fast-growing population occupied only 21% of the land, many in crowded cities and densely populated countries where sub-national governance was poor (Linard and others 2012).

3.3   Challenges of Human Rights in Development

The academic theory of path dependence argues that the decisions which we face are limited by past decisions, and critical junctures occur when existing political structures fail to respond to change, leading different actors to develop new dynamics and institutions (Sorensen 2014). Arguably the Sustainable Development Agenda was such a critical juncture, bringing together many related issues of human security and human rights, and potentially offering a rallying point for citizens to hold their governments to account (Gasper 2005; Ogata and Cels 2003). The AU's Agenda 2063 and its New Partnership for Development (NEPAD) include such aspirations as quality of life, environmental sustainability and climate resilience (Eyita-Okon 2022). Growing population pressures translate into demand for housing and livelihoods in the rapidly expanding informal areas that lie outside the state-planned urban spaces formed by colonialism.

African countries' colonial governance structures that have largely continued since independence create a gulf between the institutions, language and cultures of governance at national and international levels, on the one hand, and the mass of its peoples struggling with basic issues of survival, on the other. Governments have too often failed to provide public goods to informal settlements under the stresses of rapid urbanization. Local communities have to pool their resources to meet their needs, for instance digging boreholes and wells for water; insanitary informal settlements create health hazards, communicable diseases and untimely deaths; the use of kerosene lanterns and candles for lighting, firewood for cooking, and illegal tapping of electricity risk fire disasters; harsh living conditions increase crime and violence, as young gangs use violence to survive, and secure territory to control income-generating activities (Akanle and Adejare 2017).

An example of such governance failure is offered by Birch (2021), who investigated different responses of urban governance to disease outbreaks of ebola and cholera in the capitals of Liberia and Harare (Monrovia and Harare) respectively. In Harare degraded urban infrastructures and political competition contributed to the health crisis, turning the cholera outbreak into a 'man-made' crisis; in Monrovia the initial outbreak of ebola in remote rural areas soon spread to the capital's insanitary informal areas with poor public health provisions. Neither city had any formal political opposition that could contest the actions of the national government, while the international community obscured the role of poor governance that was allowing the outbreak to spread. Eventually new investment in water and sanitation did create better resilience, but only after many lives had been lost.

The tension between universal human rights and 'African solutions for African problems' is encountered especially for its women, who often find themselves disinherited and impoverished by patriarchal customary authorities. Female-headed households in Africa have increased because male partner deaths from disease and conflict contribute to family disruption, as well as unpartnered adolescent fertility. The Covid pandemic aggravated the existing situation, with newly widowed women often lacking family support and denied access to housing and land. Women are increasingly establishing a home without men's involvement as a 'domain of autonomy', an asset which they can let or sell. Gender inequalities and exclusion, often deriving from past colonial policies, continue to affect women's everyday lives and limit their access to land, property, and livelihood activities, while unsuitable housing designs and land-use patterns contribute to disparities (Bhatasara 2021).

Civic engagement and participation is promoted by the UN Environment Programme (UNEP) as an important component in what it calls the environmental rule of law (UNEP 2023). Community-based action can come from a sense of place identity and attachment (Basile and Ehlenz 2020). One example is the Soweto East Project in Kibera (Nairobi, Kenya), which has been called the largest slum or informal settlement in Africa. A partnership between the Kenyan Government, UN-Habitat and the *Maji na Ufanisi* CBO (which means in Swahili 'water and development') achieved through community engagement the building of an access road, a community resource centre, better water supply, sanitation and waste management (Meredith 2021). Another UN-Habitat project, in Sierra Leone, developed voluntary relocation guidelines for sustainable slum resettlement (Sait 2021). Such projects, however, need support from outside institutions to achieve their potential.

3.4 Issues of Land and Access to Land

Governance operates in physical space and translates into material realities on the ground. The stakeholders in land governance comprise state and non-state actors, private, corporate and professional interests, and the public. Many developing countries struggle with land-related conflicts that hinder their development (Wehrmann 2019). The Peruvian economist Hernando de Soto claimed a framework of secure, transparent and enforceable property rights to be a critical precondition for poverty reduction and economic growth—even the solution to global poverty (De Soto 2000). Land can be seen as the single greatest basic resource in a country, which makes good land governance vital to achieving good governance in general. Access to land, security of tenure and land management touch all aspects of how people live and earn a living. Land-based taxes can raise revenue for public finances, and title registration provide security for people and mortgage transactions (Franzsen and McCluskey 2016). Weak governance means that the poor are not protected, left marginalised and outside the law, and land is not best used to create wealth for the benefit of society. Land governance is thus about the policies, processes and institutions by which land, property and natural resources are managed. It is relatively new in international development policy, introduced by FAO and the World Bank in the early twenty-first century as an extension of the concept of 'land management' to include also aspects of governance and the political economy of land, the manner in which decisions about access to land and its use are implemented and enforced, and how competing interests in land are managed (UNEP 2023).

The so-called and much-debated 'land question' in Africa usually refers to the exclusion of most of the population from getting access to land, whether by a white settler minority, post-colonial elites or foreign investors. After independence many African governments reformed their land and planning laws in attempts to redress colonial injustices, but kept control in the hands of the state. These reforms have not always helped to achieve broad-based socio-economic development, as powerful vested interests benefited from an environment of insecure land rights through corrupt and fraudulent land allocations (Boone 2014; Manji 2012; Onoma 2010). Such abuses are gradually becoming less acceptable: an AU anti-corruption convention has existed since 2003), and some lands ministers and officials have been convicted in recent years.

The African continent has the highest proportion of customary land in the world (an estimated two-thirds of the usable land area), and millions of its people live in informal settlements on customary land that are illegal in the eyes of the state, at risk of arbitrary demolition and eviction. Indigenous African cultures saw land as a bundle of social responsibilities, very different from the Western view of it as conferring a 'bundle of rights' upon the owner, and treating private property as superior to communal systems (Home 2013). Land law reforms are now asserting an equivalent legal status for customary and 'modern' property rights, but the two tenure types are often in competition. Customary land tenure and traditional authorities are proving tenacious in supporting collective or communal rights, and the AU has created a Forum of African Traditional Authorities in part to support them. Official disapproval and harassment of informal settlements, which are seen as illegal squatters deserving forced removal, have deep roots in the colonial experience, and have only moderated in recent years as slum-dwellers exert pressure on their elected politicians. Large-scale evictions by government are still being justified as necessary planning interventions, for example in Nigeria (Lagos and Abuja), Zimbabwe and elsewhere (UN-HABITAT 2007).

The AU in 2009 adopted a declaration on land (AU 2009), and an African Land Policy Centre in 2014 was formed in Addis Ababa as a joint programme by the AUC, the African Development Bank and the United Nations Economic Commission for Africa. Its land policy guidance (AUC 2012) recommended reducing the 'overwhelming presence of the state in land matters', but that depends upon the member states being willing to give up some power and having the capacity to implement new land policies. The AU has also created a Network of Excellence in Land Governance for Africa (NELGA) to build capacity in higher education, The linguistic legacies of colonialism, and the AU's origins in anti-colonial struggles, affected the choice of location for NELGA's regional 'nodes' of operation, with Anglophone and Francophone countries maintaining separate nodes in the West

African region: in Senegal for the Francophone participants, and in Ghana for the Anglophone (Home 2021a).

UN-Habitat, with its mission of 'a better quality of life for all in an urbanizing world', established in 2006 the Global Land Tools Network (GLTN). This claims to be a 'dynamic and multisectoral alliance of international partners committed to increasing access to land and tenure security for all, with a particular focus on the poor, women and youth'. The GLTN Strategy for the Sustainable Development period reviewed the international frameworks for land governance, and enlisted 80 partner organizations in four collaborative cluster groups: international civil societies (both urban and rural), training/research institutions, and professional bodies (GLTN 2019). Other key commitments are tenure security for all, enabling responsible land governance and sustainable land use, and generating land-based revenues (Franzsen and McCluskey 2016).

The GLTN has been developing and testing at country level some twenty land tools, of which two in particular are relevant for Africa. Firstly, the Social Tenure Domain Model (STDM) offers a standard for representing people–land relationships independent of levels of formality, legality and technical accuracy (Lemmen 2013). Secondly, Participatory and Inclusive Land Readjustment (PILaR) seeks to supplement the pre-existing land readjustment model with a more inclusive negotiation process. Land-ownerships are pooled to plan urban extensions, increase densities, and finance better physical infrastructure, public space and other amenities. This can narrow the gap between centralised control approaches and the reality of people's needs on the ground, so that costs and benefits may be better shared between landowners and other stakeholders in a less confrontational approach than compulsory expropriation (Chavunduka 2021; UN-Habitat 2016).

As well as the GLTN's land tools, the international surveying community through its International Federation of Surveyors (FIG) promotes for developing countries basic and inclusive land administration systems: a Framework for Effective Land Administration (FELA) and the Fit-For-Purpose Land Administration (FFPLA) approach (Enemark 2021, 2022). Land professionals such as lawyers, surveyors and planners still remain the custodians of the land administration systems, and political will is needed to show the longer-term benefits to society in implementing the new system (Home 2021b). International moves towards FFPLA involve experimenting with digital technologies, offering online land registration and transactions (Enemark 2021, 2022).

3.5 Towards SDG 11 (Sustainable Cities)

Probably the two greatest challenges now for African local government and development are rapid urban growth and climate change. The SDGs adopted by the UN international community in 2015 gave this new urgency, specifically SDG11 ('Inclusive, safe, resilient and sustainable cities') and SDG13 ('climate action'). In 2016 the Habitat III Conference in Quito (Ecuador) also adopted a wide-ranging New Urban

Agenda (NUA 2016), and the UN-Habitat's Executive Director has claimed SDG11 to be 'at the heart of the SDGs' (UN-Habitat 2019a, b, c).

To take SDG11 first, the world's urban population now exceeds that of rural areas, a hugely significant and almost certainly irreversible historical shift. Increasingly such urban populations live in megacities of over ten million inhabitants, of which Africa already has several in Lagos, Cairo and Kinshasa (Rukmana 2020). Research from satellite mapping by Linard and others (2012) found that in 2010 some 90% of Africa's population occupied only 21% of its land area, many of them living in crowded cities and densely populated countries, while the other 10% were scattered over the rest of the land, much of which was deserts and mountains barely inhabitable. There seems to have been some decline in Africa's urban footprint and growth rates in recent years (exemplified by. South Africa and Zimbabwe), but the largest cities continue to grow faster than smaller ones. High urban densities occur in the poorest countries (exemplified by the DRC and Mozambique), taking the form of overcrowded informal settlements rather than liveable and productive places (Gambe 2023). UN-Habitat research has recommended that a desirable urban density would be 15,000 people per square kilometer (150 per hectare) for 'a new and sustainable relationship between urban dwellers and urban space'. This is a much higher density than the low-density car-dependent urban forms found in wealthy countries, and has implications for both land use planning and climate action (UN-Habitat 2019b).

The former British and French African colonies urban areas have differing settlement patterns as a result of official policies. British policies, notably associated with Lugard's dual mandate, sought to maintain a physical separation between the incoming colonisers and the indigenous populations, which was carried forward as apartheid policy in South Africa and elsewhere (Dubow 1989). Such policies saw towns as essentially European creations, where land was claimed by the state and then subdivided for leasing out, while Africans were discouraged by pass laws and other controls from living in towns. Excluded and banished to a 'septic fringe' of peri-urban settlements, they lived in temporary structures, perhaps under threat of demolition and displacement by the British colonial authorities (Home 2012). British indirect rule and dual mandate policies resulted in urban development with less overall plan or coordination than was found with French direct rule, which by contrast featured centralised city planning and land allocation mechanisms. Another research study that analyzed satellite sensed data showed that cities of Anglophone colonial origin have less intense land use than Francophone ones, with more irregular layouts and more 'leap-frog' development at the urban edge, configurations which make the recommended UN-Habitat densities difficult to achieve (Baruah 2021). Outside the towns and white settler lands of British colonies, customary tenure was maintained by a policy of tribal reserves or tribal trust lands, but the colonial state could take land there without paying compensation (the legal term was 'set aside'), if required for what it deemed the 'public interest', such as mining, forestry or township creation (Home 2021a).

One consequence of such British African colonial practice is that households in Anglophone informal areas now have poorer connections than Francophone ones to electricity and piped water, linked to the difficulties and higher costs of providing infrastructure for urban sprawl (Baruah 2021). The political settlement at independence has facilitated the phenomenon that Watson (2014) has called 'African urban fantasies'. These are planned and self-contained satellite settlements around major cities, funded largely by international real estate investment for an upper-income market (Goodfellow 2017). They are far removed from the impoverished masses, who are often evicted to make way for these new entities, undermining the socially inclusive goals of contemporary global development agendas. Some forty such new visions of self-styled global, smart and sustainable cities were identified in Africa by Abubakar (2021).

In recent years the concept of urban law, both national and local, has gained attention from academics concerned with the New Urban Agenda (NUA) and SDG11 aspirations: to 'make cities and human settlements inclusive, safe, resilient and sustainable' (NUA 2016). The NUA focuses on tenure security, housing and community development as key requirements for social, economic and ecological sustainability, allowing access by all to adequate, safe and affordable housing, green and public spaces, transport systems and basic services such as water, food, electricity, sanitation, and waste deposal (Davidson and Tewari 2019). This new urban law implies law reforms that can bring together planning and housing regulations, and acknowledge that urban citizens have a 'right to the city' (Njoh 2017). Such reforms can be difficult to achieve against opposition from vested interests (Glasser and Berrisford 2015; Manji 2012; Berrisford and McAuslan 2017), and UN-Habitat research has found urban planning in Africa to be largely ineffective when measured against various indicators (UN-Habitat 2019a). In Kumasi (Ghana) research found that incremental housing development did not conform to regulations because of excessive bureaucratic formality, builders inadequately aware of them or ignoring them, and insufficient enforcement; it recommended improving public awareness and homeowner attitudes (Asibey 2023).

3.6   Towards SDG13 (Climate Action)

Africa is the continent that contributes the least to global greenhouse gas emissions, yet faces some of climate change's harshest consequences. International and regional responses to SDG13 should be reinforcing urban planning and management, but progress has been slow, even in the 2020s that were identified as the 'decade for action' (UN-Habitat 2019a). An African country (Egypt) hosted the Conferences of the Parties (called COP) to the United Nations Framework Convention on Climate Change in 2022, and urged drastic response for cities to cope with the risks of climate change. SDG15 ('Protect, restore and promote sustainable use of terrestrial ecosystems') poses a challenge for local land governance, which remains bureaucratic and vulnerable to exploitation and corruption, while struggling with greatly increased populations and demands. Climate action needs legislative and perhaps even constitutional changes within the countries signed up to it, yet few states have achieved the SDG target of formulating a framework climate change law, notwithstanding UNEP efforts in that area (UNEP 2019). At local level individual cities, often struggling with political instability and conflict, show little capacity or appetite for climate change action, and investment continues to concentrate on high-income real estate and global business competitiveness rather than attempt to address intractable longer-term issues such as climate action. A research study of Gizeh, third largest city in Egypt with World Heritage listed pyramids, found that the centralisation of decision-making at national level and under-resourcing of local government meant that city officials knew little about climate action, policies and their responsibilities, even though Egypt was a signatory to the SDGs and indeed hosted the COP 27 Climate Change conference in 2022 (Yarra 2021). Thus climate action policies for cities have been slow to gain traction despite being on the international agenda for the last 30 years (Zaheer Allam 2022).

Climate action involves local responses to the disasters that are occurring with increasing frequency. The internationally agreed Sendai Framework for Disaster Risk Reduction 2015–2030 concerned physical adaptation measures, early warning systems, and emergency responses to reduce impacts (Pearson and Pelling 2015). The Sendai approach is now expanding into wider strategies for urban resilience that include basic infrastructure and services, poverty reduction, and social exclusion (Borie 2019). Urban resilience is broadly defined as the capacity of urban individuals, communities, institutions, and systems to survive, adapt, and grow, no matter what chronic stresses and acute shocks come. This more holistic and pro-active approach sees cities as complex adaptive systems or networks that interact with political and institutional processes, with urban resilience and risk management involving an interplay between complex development processes, mapping vulnerable settlements in high-risk zones, achieving disaster-resilient building construction, and a governing strategy of learning and adaptation (Home 2021a).

The COVID-19 pandemic that started in 2020 posed brutal risks to the billion people globally who live in poor and densely populated urban areas: lockdowns were almost impossible when basics of water and soap were unavailable for handwashing, people had to leave their homes to work for daily survival needs, while stay-at-home orders put women at greater risk of violence (UN-Habitat 2020). The pandemic provided an insight into the tasks needed for city planning and management regimes in facing climate action, yet these regimes are often disinterested in issues of disaster risk and resilience building until the disasters actually happen (Adeleye and Ajobiewe 2022).

As well as COVID-19 the urgent need for climate action is increased by flood risk from unregulated informal settlements in flood-prone areas (Andreasen 2022; Anwana and Oluwatobi 2023). Most occupiers acquired land informally when moving into the city, through social connections, brokers, and obtaining verbal or written transfer agreements from ward officials, while the few who inherited land did not use the legal transfer processes. Initially the incomers put up temporary dwellings, and gradually building structures emerged. Mapping of flood-prone areas may not exist, and if they do enforcement measures may not work without community support (Kemwita 2022; Oteng-Ababio 2022).

Climate-related extreme weather means not only flood risk but also more intense, frequent and long-lasting hot spells, which are particularly dangerous for older residents and those living in poor informal settlements without air conditioning or the cooling benefits of green space or water. Such heat-related impacts can be reduced if local governments are adequately prepared, coordinate action, invest in public infrastructure, and improve early-warning systems. Examples of action are resilience hubs where people can cool down during heatwaves, and basic measures like bus stops being shaded.

In Africa Sierra Leone and Guinea Bissau were listed as high on the IPCC's 2016 list of countries most vulnerable to climate change. In Freetown, the capital and largest city in Sierra Leone, extreme heat has increased over the past decade, and a third of the population live in informal settlements, where houses are densely built of temporary materials that trap heat, and in disaster-prone areas like the seafront or hillsides. The city in 2021 appointed a chief heat officer, tasked with raising public awareness about extreme heat, improving responses, and collecting, analysing and visualising heat impact data for the city. Her recommendations for improvement include such matters as urban greening and parks, community health initiatives, early warning from weather forecasts, and knowledge transfer through inter-city collaboration (Adegun 2023).

The climate change connection with urban law and governance reinforces the need for improvements to informal settlements, which require attention in theory and practice (Finn and Cobbinah 2023). After decades of government evictions and demolition, communities are increasingly demanding greater accountability and transparency from government, and support to empower their improvement efforts. The informal areas may lack basic services, and be difficult to navigate physically, with inadequate street and property addressing, poor road and path networks, and are still seen as inferior to 'formal' developments. SDG16 and the NUA aspire toward 'peaceful, inclusive and participatory societies', and SDG 17 promotes partnerships between stakeholders. Some local projects in informal settlements are identifying appropriate opportunities, (Finn and Cobbinah 2023), and civil society can lead through mobilizing new-style social movements (Lemanski 2019). New attitudes towards tenure security have softened official hostility to squatter settlements, as governments recognize the political costs of eviction, and tenure regularisation can be linked to physical upgrading measures. Kenya (where UN-Habitat has its head-quarters) has various community-based organizations negotiating with government to secure access to land, shelter and basic services for the urban poor (Muchadenyika and Waiswa 2018).

Air pollution is yet another consequence of climate change. In Nigeria air pollution from oil and gas exploration activities and gas flaring have created serious environmental problems that add to global warming and climate change, yet its government has shown inadequate commitment to various international agreements on climate change (Gasu 2022). The concept of climate justice can be applied to improve access to justice and protect climate change victims, with recent reforms and initiatives by the Nigerian government perhaps offering progress towards climate change litigation (Bouwer 2022; Ekhator and Okumagba 2023).

Some cities cope with crises and maintain their resiliency, as the city of Gondar (Ethiopia) has shown through five specific development strategies: strong leadership; applying principles of good governance in planning and development; building a strong regional economy from the rich agricultural resources of the region; developing a strong tourism sector; and building infrastructure in transportation, water, power, and other basic services to entice private investment (Wubneh 2022).

3.7 Conclusions and Future Directions

The development challenges facing Africa have been adversely affected by the impact of the Covid-19 pandemic and accelerating climate change, with extreme weather events such as flooding and drought increasing in frequency. Also poverty, inflation and corruption are negative forces holding back the continent's development. This chapter has sought to show the complex processes and constraints upon resources which hamper attempts to tackle climate change challenges, while basic needs and human rights are still inadequately met. Among the possibilities for legal reform are constitutional changes to redistribute statutory responsibilities and powers between levels of government, under principles of subsidiarity and through regional co-operation. Civic engagement also needs developing to provide urban resilience, disaster risk management, and service delivery.

The political settlement at independence in African countries often continued wealth inequalities and centralized governance, and is now increasingly challenged by population pressures, rapid urbanization and climate change, while the global community's expectations have grown, as expressed in the SDGs. The so-called youth bulge in Africa's demography puts new demands upon the governing class, while existential threats grow. Dysfunctional national land laws and administration are increasingly seen as a major economic obstacle to African development, and there is no easy route to improve Africa's local governance: resources, political leadership, law reform, and investment in systems and technologies are all needed.

In spite of these negatives, there are reasons for optimism, with much having been achieved within a few years, at least to better understand the problems and frame solutions. Local action can influence the needed transitions in governance and social behaviour through involvement by young people, civil society organizations, the private sector, academia, and other stakeholders. Techniques and political skills are developing in mediation, dispute resolution and local coalition-building, while academic scholarship is adding new knowledge through interdisciplinary approaches and collaborative partnerships. Theory and practice is being rethought to change negative colonial legacies by reform of laws and regulations, improve public space both physical and figurative, stimulate entrepreneurship and innovation, and build stronger civic society. Technological advances in remote sensing and unmanned drones are improving land governance, environmental resource management and land use planning. Advances in survey technology championed by younger surveyors are adding more democratic data-capture techniques, and digital land administration offers the potential of better registration and protection of property rights.

Land governance at a local everyday level means communities and neighbourhoods negotiating formal and informal rules, which seems to be occurring across Africa, and new legal structures for land management are developing, such as community land development trusts and innovative land tools. Traditional authorities are proving resilient in the political landscape, and helping better community participation in development efforts. New knowledge and actor networks are developing new thinking and connections, such as FIG, the GLTN partner group and NELGA.

References

Abubakar IR (2021) Governance challenges in African urban fantasies. In Home (ed) op. cit.155–170

Adegun O (2023) Africa's first heat officer, The Conversation, 23 February

Adeleye O, Ajobiewe T (2022) Climate change, COVID-19 and war: triad litmus test questioning the conscientiousness for collective action. Town Reg Plann 81:1–6

Akanle O, Adejare GS (2017) Conceptualising megacities and megaslums in Lagos, Nigeria. Afr Public Serv Deliv Perform Rev 5(1):9

Zaheer Allam (ed) & others (2022) Cities and climate change: climate policy, Economic Resilience, Palgrave Macmillan

Andreasen MH, others (2022) Built-in flood risk: the intertwinement of flood risk and unregulated urban expansion in African cities. Urban Forum. https://doi.org/10.1007/s12132-022-09478-4

Anwana EO, Oluwatobi MO (2023) Analysis of flooding vulnerability in informal settlements literature: mapping and research agenda. Soc Sci 12:40

Asibey MO, others (2023) Incremental housing and compliance with development control in urban Ghana. J Urban Aff. https://doi.org/10.1080/07352166.2022.2160337

AU (2009) Declaration on land issues and challenges in Africa. Assembly/AU/Decl.l (XIII) Rev.I. 2009. AUC-ECA-AfDB Addis Ababa

AUC (2012) Tracking progress in land policy formulation and implementation in Africa. Addis Ababa

AUC (2015) Agenda 2063: the Africa we want. Addis Ababa

Baldwin K, Holzinger K (2019) Traditional political institutions and democracy: reassessing their compatibility and accountability. Comp Pol Stud 52(12):1747–1774

Baruah J, others (2021) Colonial legacies: shaping African cities. J Econ Geogr 21(1):29–65

Basile P, Ehlenz MM (2020) Examining responses to informality in the global south: a framework for community land trusts and informal settlements. Habitat Int 96

Berrisford S, McAuslan P (2017) Reforming urban Laws in Africa: a practical guide. African Centre for Cities, Cape Town

Bevir M, Rhodes RAW (2010) The state as cultural practice. Oxford University Press, Oxford

Bhatasara S (2021) Women, land and urban governance in colonial and post-colonial Zimbabwe. In Home R op cit 207–224

Birch H (2021) Urban governance and disease outbreaks: cholera in Harare and Ebola in Monrovia. In Home R op cit 299–316

Boone C (2014) Property and political order in Africa. Cambridge University Press, Cambridge, UK

Borie M & others (2019) Mapping narratives of urban resilience in the global south. Glob Environ Chang, 54: 203–213

Bouwer K (2022) The influence of human rights on climate litigation in Africa. J Hum Rights Environ 13(1):157–177

Chavunduka C & others (2021) Stocktaking participatory and inclusive land readjustment in Africa. In Home R op cit 137–154

Cirolia LR (2021) Financing African cities: a fiscal lens on urban governance. In Home R op cit 35–52

Conklin, J (2001) Wicked problems and social complexity. Available online: www.cognexus.org

Davidson N, Tewari G (eds) (2019) Global perspectives in urban law: the legal power of cities. Routledge, London

De Soto H (2000) The mystery of capital: why capitalism triumphs in the west and fails everywhere Else. Black Swan, London

De Visser J (2021) Food security, Urban Governance and Multilevel Government in Africa In Home op cit 269–280

Dubow S (1989) Racial segregation and the origins of apartheid in South Africa 1919–36. Macmillan, Oxford

Ekhator E, Okumagba EO (2023) Climate change, multinationals and human rights in Nigeria: a case for climate justice. Forthcoming. In: Bouwer K, others (eds) Pursuit of climate justice in Africa. Bristol University Press

Enemark S & others (2021) Fit-for-purpose land administration—providing secure land rights at scale, Land 10(9) 972

Enemark S (2022) Responsible land governance and secure land rights in support of the 2030 global agenda. FIG Congress, Warsaw

Eyita-Okon E (2022) Urbanization and human security in post-colonial Africa. Frontiers in Sustainable Cities 4:917764

Finn BM, Cobbinah PB (2023) African urbanisation at the confluence of informality and climate change. Urban Stud 60(3):405–424

Franzsen R, McCluskey W (eds) (2016) Property tax in Africa: status, challenges, and prospects. Lincoln Institute of Land Policy, Cambridge, MA

Fukuyama F (2013) What is governance? Governance 26(3):347–368

Gambe TR & others (2023) The trajectories of urbanisation in southern Africa: a comparative analysis, Habitat Int, 132, 102747

Garcia Pena MC (2023) Planning and evaluation of the 2030 agenda: governance, localization and balance. University of Malaga

Gasper D (2005) Securing humanity: situating 'human security' as concept and discourse. J Hum Dev 6(2):221–245

Gasu M, others (2022) International and national policy responses to combating global warming and climate change in Nigeria. Town Reg Plann 81:113–123

Glasser M, Berrisford M (2015) Urban law: a key to accountable urban government and effective Urban Service delivery. World Bank Legal Rev 6:209

Global 2030 Agenda (2015) Transforming our world: the 2030 agenda for sustainable development. United Nations, New York

GLTN (2012) Handling land: innovative tools for land governance and secure tenure. UN-Habitat, Nairobi

GLTN (2019) Strategy 2018–2030. UN-Habitat, Nairobi

Goodfellow T (2017) Seeing political settlements through the City. Dev Chang:1–24

Goodfellow T (2017b) Urban fortunes and skeleton cityscapes: real estate and late urbanization in Kigali and Addis Ababa. Int J Urban Reg Res 41(5):786–803

Hammond-Tooke WD (1997) Imperfect interpreters: South Africa's anthropologists, 1920–1990. Witwatersrand University Press, Johannesburg

Home RK (2012) Colonial township Laws and Urban governance in Kenya. J Afr Law 56:175–193

Home RK (2013) 'Culturally unsuited to property rights?': colonial land Laws and African societies. J Law Soc 40(3):403–419

Home RK (2021a) Urban law and resilience challenges of climate change for the MENA region. In: Olawuyi D (ed) Climate change law and policy in the Middle East and North Africa region. Routledge, London, pp 153–168

Home R (2021b) History and prospects for African land governance: institutions, technology and 'land rights for all'. Land 10:292

Home R (ed) (2021c) Land issues for urban governance in Sub-Saharan Africa. Springer. [References to chapters in this book are cited as 'in Home, op.cit.; 2021']

Kaufmann D & others (1999) Governance Matters SSRN: https://ssrn.com/abstract=188568

Kemwita E, others (2022) Land acquisition processes in flood-prone informal settlements in Dar es Salaam: a rhetoric revealing land governance. Town Reg Plann 81:67–83

Lemanski C (ed) (2019) Citizenship and infrastructure. Routledge, London

Lemmen C (2013) The social tenure domain model: a pro-poor land tool. FIG Publication 52, Eilat

Linard C, others (2012) Population distribution, settlement patterns and accessibility across Africa in 2010. PLoS One 7(2):e31743

Mahoney J, Thelen K (eds) (2010) Explaining institutional change: agency, ambiguity and power. Cambridge University Press, Cambridge

Mamdani M (1996) Citizen and subject: contemporary Africa and the legacy of late colonialism. Princeton University Press, Princeton

Manji A (2012) The grabbed state: lawyers, politics and public land in Kenya. J Mod Afr Stud 50(3):467

Mayer B (2018) International law on climate change. Cambridge University Press, Cambridge, MA

Meredith T & others (2021) Partnerships for successes in slum upgrading: local governance and social change in Kibera, Nairobi In Home R 237–256

Muchadenyika D, Waiswa J (2018) Policy, politics and leadership in slum upgrading: a comparative analysis of Harare and Kampala. Cities 82:58–67

Munshifwa EK (2019) Adaptive resistance amidst planning and administrative failure: the story of an informal settlement in the city of Kitwe, Zambia. Town Reg Plan 75: 66–76

New Urban Agenda (NUA) (2016) Habitat III United Nations. A/RES/71/256

Njoh A (2017) The right-to-the-city question and indigenous urban populations in capital cities. J Asian Afr Stud 52(2):188

Ogata S, Cels J (2003) Human security: protecting and empowering the people. Glob Gov 9(3): 273–282

Onoma AK (2010) The politics of property rights institutions in Africa. Cambridge University Press

Oteng-Ababio M, others (2022) Built-in flood risk: the intertwinement of flood risk and unregulated urban expansion in African cities. Urban Forum. https://doi.org/10.1007/s12132-022-09478-4

Parnell S, Oldfield S (eds) (2014) Routledge handbook on cities of the global south. Routledge, London

Pearson L, Pelling M (2015) The UN Sendai framework for disaster risk reduction 2015–2030: negotiation process and prospects for science and practice. J Extreme Events 2:1571001

Rotimi S (2022) De/Centralization in Nigeria, 1954–2020. Reg Fed Stud. https://doi.org/10.1080/13597566.2022.2134350

Rukmana D (ed) (2020) Routledge handbook of planning megacities in the global south. Routledge, London

Sait MS (2021) Should Monrovian communities agree to voluntary slum relocations. In Home op cit: 339–354

Shaw MN (1996) The heritage of states: the principle of Uti Possidetis juris today. Yearbook Int Law 67:75–154

Sorensen A (2014) Taking path dependence seriously: a historical institutionalist research agenda in planning history. Plan Perspect 30(1):1–22

UCLG (2019) Towards the localization of the SDGs, Barcelona

UNEP (2019) Climate Change Toolkit. Available at https://climatelawtoolkit.org/

UNEP (UN Environment Programme) (2023) Environmental rule of law: tracking Progress and charting future directions, Nairobi

UN-Habitat (2007) Forced evictions: towards solutions? Second report of the advisory group on forced evictions to the executive director of UN-Habitat. HS/932/07E

UN-Habitat (2016) Remaking the urban mosaic: participatory and inclusive land readjustment, Nairobi

UN-Habitat (2019a) Effectiveness of planning law in sub-Saharan Africa, Nairobi

UN-Habitat (2019b) A new Strategy of sustainable Neighbourhood planning: five principles, Nairobi

UN-Habitat (2019c) SDGs decade of action, Nairobi

UN-Habitat (2020) COVID-19 Response Plan, Nairobi

UN-Habitat (2023) Land and property taxation in fragile states, Nairobi

Watson V (2014) African urban fantasies: dreams or nightmares? Environ Urban 26(1):213–229

Wehrmann B (2019) Land conflicts—a practical guide to dealing with land disputes. In: Scoping and status study on land and conflict., HS/050/16E. UN-Habitat, Nairobi

Wubneh M (2022) Planning for cities in crisis: lessons from Gondar. Springer Nature, Ethiopia

Yara E, others (2021) Urban climate change governance within centralised governments: a case study of Giza, Egypt. Urban Forum. https://doi.org/10.1007/s12132-021-09441-9

Chapter 4
Territorial Development and Economic Dynamics in Morocco: What Prospects for Equitable Regional Development?

El-Hassan Farhat and Khadija Darmame

4.1 Introduction

With the socio-economic crisis that Morocco experienced after its independence, the actions of public authorities were guided by two main commitments. The first one was to multiply the number of territorial subdivisions to align the politico-administrative systems with the State's new organization. The second commitment was to enhance the role of the State in designing and implementing economic and social programs conducive to regional development. Thus, planning evolved between the socio-economic orientation of the central government and the aspirations of local governments, which were hindered by insufficient autonomy and a failure to fully acknowledge of communities' needs. In both cases, the spatial dimension was of paramount importance and since then, all short- and medium-term regional economic development strategies have been embedded in this spatial duality. This commitment could only be fulfilled within the framework of an appropriate space: initially within the province, then within a wider territory, the region (Farhat 2004).[1]

---

[1] Currently, Morocco has 75 provinces/prefectures grouped into 12 regions.

E.-H. Farhat (✉)
Polydisciplinary Faculty of Safi, Cadi Ayyad University, Marrakech, Morocco

K. Darmame
Al Akhawayn University in Ifrane, Ifrane, Morocco
e-mail: K.Darmame@aui.ma

4.2   Historical Overview of Regional Development Policies in Morocco

4.2.1   Development: Between Industrial Illusions and a Controversial Provincialization

The political and administrative upheaval initiated by independence materialized in an increase of territorial and functional divisions. These were presented as solutions to remodel an archaic national space, and to establish national sovereignty through the promotion of a certain number of deconcentrated/decentralized administrative institutions. These two types of policies were set up, in principle, to promote two diametrically opposed missions: the territorialization of socio-economic development and the maintenance of order. According to Rousset et al. (1989), this embodies the interference of the authority agent in the affairs of the decentralized provincial collectivities. The result was a legal arsenal meant to regulate the omnipotence of State agencies in local affairs, veering more towards the rigidity in the State's territorial services and the politicization of the territory.

Our understanding of the provincialization of development suggests the following: the state has strengthened the power of provinces without reshaping the unequal relationship between them and a deconcentrated and decentralized national administration. The province became the basic entity to conduct regional development policy, despite decentralized planning. In fact, regional socio-economic development was mainly orchestrated by the State's institutions, with limited participation from elected assemblies.[2] This is referred to as subordination or hierarchical oversight of all decentralized bodies. Tasked with maintaining order above all, State representatives at the local and provincial levels have exercised full authority over local elected councils and over the provincial/prefectural assemblies. Therefore, not only did the provincialization fail to yield significant results, but economic and social disparities between rich and poor provinces have worsened. Public and private investments have not been evenly generated nor have they structured the provincial space adequately as they were concentrated in the historically well-equipped provinces. "Provincialization" should have intensified the transfer of financial, human and legal resources in the service of spatial development in line with the population's expectations. However, the rigidity of this reform has become a hindrance to all such efforts.

---

[2] Despite territorial reforms and the discourse that follows, the personnel and activities of local elected assemblies are directly overseen and monitored by the appointed provincial administration and its financial controller.

4.2.2 Regionalization as an Alternative to Development

Regionalization in Morocco is gradual process that has evolved over time. At the outset of the experiment in 1971,[3] the region consisted merely of a consultative assembly with no involvement in economic development. Based on this deficiency, and following the constitutional revisions of 1992 and 1996,[4] the region became a local authority with legal personality and financial autonomy. Within the framework, territorial development was seen as crucial to achieve a better distribution of the population based on natural resources and economic activities.

The official discourse portrayed regionalization as a "trigger" for democratic dynamism and local development. An extensive development program was proposed through various investment plans and codes. A whole panoply of comparative advantages[5] and tax incentives were instituted to encourage domestic and foreign companies to establish themselves outside of the Atlantic seabord. Indeed, an industrial decentralization initiative was undertaken to promote secondary hubs such as Agadir, Béni Mellal and Meknes with the aim of alleviating congestion along the Casablanca-Kenitra axis. Yet regionalization barely impacted local development, pushing the State to rectify the situation on several occasions.[6] It proceeded with an unprecedented sharing of its power with elected representatives, delegating several powers but without the appropriate financial transfers. Neither the conventional organization nor the dynamics of the national territory were altered by the new administrative arrangements. Therefore, this constitutional weakness kept the predominance of central agencies, preventing the regions from propelling their own development.

4.2.3 Limitations of Regionalization

It appears that there has been a disconnect between rhetoric, actions and outcomes of development policies. For example, the 1999–2003 plan failed to deliver genuine solutions, despite promoting partnership among all local players, and both public

---

[3] Seven economic regions were created by a 1971 law, but these regions had neither legal personality nor financial autonomy.

[4] Sixteen new regions were created in 1997 to meet the challenges of development. Law 47/96, promulgated in 1997, conferred broad powers on the regions, predisposing them to promote economic, social, cultural and environmental development.

[5] Traditional factors, or comparative advantages, are made available to manufacturers to boost the location of economic activities. According to R. Kahn, these factors are characteristic of developing countries. They include labor and raw material costs. These factors are reflected, for example, in the composition of the BERI (Business Economic Risk Index), which measures the business climate in 50 countries.

[6] In particular, the 1976 Communal Charter which revolutionized local decentralization by transferring a number of powers to elected officials and communal representatives.

and private capital continued to be invested in historically attractive industrial regions. Therefore, private investment was feebly impacted by development policy, as it did not create the necessary conditions to meet industrial demand and aspirations in all regions. In essence, the investment codes implemented under regionalization did not engender any qualitative advantages and had only a limited impact on industrial decentralization.

Consequently, regionalization did not change the centralized nature of the development plan. Decentralized structures were neither effectively involved in the design of national development programs nor in their own development (Brousky 1988). As a result, national industrial development is still governed by structural and functional dependence on the Casablanca metropolis. Occasionally, certain industrial zones have benefited from the economic externalities of Casablanca, and these have been sometimes subject to the directives of the investment codes, enabling these zones to develop only quantitative advantages.[7] This industrialization mechanism led to a limited and unbalanced remodelling of industrial space, which has remained fixed solely to the needs of the Casablanca pole.

These comparative advantages did not develop a territorial competitiveness that would enable appropriately decentralized industrial development. One of the main factors is that industrial development depended on such traditional location factors as the cost of labour (low production cost), the cost of productivity, level of taxation (contribution of investment codes), and geographical proximity. It was believed that the guidelines of the 1999–2003 plan alone were sufficient to create the necessary conditions for local development. This did not happen. Therefore, in the pursuit of a more equitable distribution of investment across the national territory, it was opportune to continue long-term planning, culminating in the industrial Emergence Plan in 2009, which recommended development by 2020 of numerous sectors of the economy, such as the Halieutis Plan for the sea and the Green Plan for agriculture.

4.3   Advanced Regionalization: The Ultimate Framework

From then on, there has been a lack of understanding of decentralized development.[8] Development plans have focused solely on resource exploitation to meet national and international needs, relying solely on sectoral trial and error, underestimating the spatial dimension. Whereas local development requires that territories be considered as the basis of local livelihoods where living conditions and the right to development are legitimate, the territory was forgotten in the elaboration of national public policies. Additionally, local development calls for the harnessing of latent energies

---

[7] Such as more or less well-equipped industrial zones, locational advantages of proximity to the Atlantic Seaboard.

[8] Despite previous reforms, the territory is still not taken into account to increase non-market exchanges and cooperation between economic agents.

and resources, as well as genuine decentralization. The bottom-up approach to development is still relevant, and no new planning mechanisms have been formulated for the benefit of local authorities. It is certain that Morocco's transition from a nascent form of regionalization, subject to state tutelage, to a more advanced one will strengthen democracy, a prerequisite for local governance.

An alternative, therefore, emerges through the advanced regionalization process launched in 2015.[9] It becomes evident that the foundation of this decentralized regional development depends on collective action at the local level. The role of local and regional authorities in ensuring the region's competitiveness in terms of investment is of prime importance. The actions of local authorities must embody this new decentralization, taking advantage of its progressive nature[10] to forge regional development. A dialectical relationship is therefore emerging between companies and the region, which presupposes that companies receive regular support at all levels, and that local authorities act to make the region attractive. By becoming a brand image for companies, the territory becomes an essential variable and stake in industrial location. It is not only "the space where economic activities are located" but gradually a component of the production process of companies, an immobile factor of production (Kahn 1998).

This alternative is essential, given that development is a collective action between local players and the state (Perrin 1983). It is therefore certain that the hegemony of the state in development matters will weaken in favor of local initiatives. The combination of the new decentralization and the globalization of the Moroccan economy will shape this approach which considers territory[11] as a source of development by capitalizing on the value of its terroirs. The territory thus asserts itself as an elementary unit for this basic development. Local authorities tend to think of their territory as a product that needs to be enhanced and promoted, particularly to investors. (Kahn 1998) In the past few decades, a considerable body of literature has developed listing numerous semantic difficulties. Sometimes we speak of endogenous (self-centred) development. Friedmann (2022) calls it agropolitan, or

---

[9]The 2011 constitutional revision which advocated a number of reforms, including advanced regionalization, was launched in 2015.

[10]The region has obtained real competencies within the framework of advanced regionalization: direct election by the population of both the president of the assembly and regional councillors. Regions draft their own development plans and are responsible for implementing them. The president of the region is the authorizing officer of the regional budget.

[11]Territory is a polysemous concept. According to Guy Diméo, "Territory is the economic, ideological and political (i.e., social) appropriation of space by groups who give themselves a particular representation of themselves and their history" (Les territoires du quotidien, 1996, p. 40). For Jacques Lévy,"territory has two dimensions: a material, geographical nature in the true sense of the term, and an ideological or ideal content." The author insists on the interdependence between social, spatial and cultural facts, taking the example of the different interpretations of the same Sudano-Sahelian space in Niger by culturally different social groups. C Courlet et al. (2006) come to the same conclusion as Diméo. Whether it's called an industrial district or a localized production system, the territory is more a support for social relations than a predetermined spatial framework.

bottom-up development.[12] However, it is an integrated form of development that takes into account social, cultural and agro-industrial aspects. It goes hand-in-hand with flexible consultation, in opposition to the old managerial methods, and with self-regulating based on the politico-legal conditions of its spatial units, especially the region.

The emergence of this new spatial thinking (Dumas 1989),[13] conveyed by broader regionalization, calls into question the strategies Moroccan has applied. Beyond the reorganization of regional development, what does this new spatial thinking propose? To ensure the renewal or conversion of the productive system, can we imagine a synergy between scientific research and regional industry? And to what extent can local authorities create or prepare the attractiveness of the region to carry out this economic conversion?

4.4 The Localization of High-Tech Industry Is Essential for Decentralized Development

4.4.1 Which Localized Production System Is Best Suited to Technological Innovation?

We can begin to question the approaches to regional development and the possible modalities around which this new industrial dynamic can be constructed. Will high-tech industries be located in districts, business parks, innovation centres, science parks or incubators? Despite this semantic difficulty, in all cases, these spatial forms signify a clustering of businesses, particularly of Small and Medium-sized Enterprises (SMEs), outside of traditional industrial zones, known in recent years in Morocco as technopoles or free zones. Beyond their media or modernist appeal, do these types of spaces reflect localized territorial competitiveness in any given area? Or will this new spatial dynamic be based on the regional territory's capacity to innovate and endorse technical change? Lastly, is the region becoming a key factor in industrial dynamics, mobilizing local energies to ensure industrial development? In any case, we are not looking for a single approach or an ideal spatial form on which regional industry should be organized. On the contrary, believing in the sectoral and geographical diversity of regional-local development, we aim to analyse it in order to propose possible types of industrial and scientific organization.

---

[12] As opposed to top-down development. The Moroccan experience is inspired by this bottom-up paradigm, by basing its development strategy on local authorities.

[13] Dumas speaks of cross-fertilization between research and industry. At the same time, he speaks of a "utopian ambiguity". He studied the emergence of French technopole policy at a time when the country was embarking on a major territorial reform with decentralization.

In this regard, economists and geographers have developed a number of approaches based on the design of a local flexible industrial system. They help us to better understand the foundations of local development. The emergence of one or other of these approaches is linked to the decline of the Fordist model (Dagri 2000),[14] according to which the competitiveness of the industrial production system was based on physical proximity, a necessary condition for the survival of companies and external economies (Planque 1981).

For the sake of convenience to the nature of industrial space, we have chosen the approach that considers the territory[15] as an "innovative environment" and the incubator as a territorialized model of industrial development. In contrast to the Coase-Williamson-Scott paradigm, which asserts that "industrial organization arbitrates between the firm's internal organization costs and the transaction costs between firms," Aydalot, Maillat and Perrin are among the leading proponents of this approach. For Perrin (1983), the territory is an innovative environment. Although influenced by the evolutionary theory of innovation and the regulation school, these authors are primarily concerned with determining the external conditions necessary for the birth of the firm and the adoption of innovation. There are therefore two underlying currents.

1. The first considers that the existence of an environment is a prerequisite for any economic action. In other words, business creation depends on the competitiveness of the region, which is why its followers attach particular importance to the environment. At the same time, they profess that innovation belongs to the environment. In other words, they subscribe to the "development theory of environments" and reject the theory that suggests that company location is a prerequisite for socio-economic development.
2. The second is more concerned with finding a new territorial organization for industrial development. In this respect, authors like Maillat and Perrin proposed strategies to overcome the failures of the Fordist model.

4.4.2 The Location of High-Tech Industry Is Necessary for Decentralized Industrial Development

Under which territorial model or spatial form can Moroccan regional industry be organized? In all cases, the choice of a territorialized model for the production system is based on the progress of the current territorial decentralization. This

---

[14] Dagri argues that these approaches essentially deny the decline of the Fordist mode of development, the emergence of post-Fordist production models, the dynamics of industrial districts, "high-tech" spaces, the growing importance of service activities in urban economies and their location dynamics.

[15] Zaoual (1996) develops a theory of symbolic sites. This theory is totally in line with the economics of territory. For him, a site is above all a territory that produces meaning.

enables local authorities to stimulate the intrinsic capacities of the territory and to act on the potential of scientific and human affinities capable of initiating a regional industry. Given Morocco's unique conditions and challenges, a number of specific governance models from around the world will be discussed in this chapter. We will explore the potential of these experiences, particularly the successful implementation of the Italian model, to in order to understand Morocco's specific needs and objectives in fostering innovation and economic development. We aim to provide a tailored assessment of the applicability of these models to Morocco's context, ultimately facilitating a more effective and contextually relevant policy formulation.

According to the Italian model, in developed countries the industrial district was the scale for theorizing the crisis in the Western economy. It was first used by Alfred Marshall (1898) at the end of the nineteenth century. Since then, several models, such as innovative communities, have emerged, whether or not they identify with the district, in order to resolve the crisis in industry. However, the disagreement between these milieus and the district remains fundamental. In fact, the industrial district is based solely on the existence of a group of SMEs, to the exclusion of large firms which specialize in a given sector. In addition to the reciprocal exchange of free services, these companies maintain market relations based on supply and demand. These companies carry with them the productive tradition of the geographical area concerned. In this respect, Peyrache (1992) insists that "the district's sectoral specialization is based on the know-how of local economies and on the rooting of productive forms in a territory whose origins are rural or artisanal".

The success of Italian industrial districts owes much to the role played by political and regional authorities, both in terms of cooperation and of managing competition between companies. On the other hand, the specialization of companies in a common branch or even around a particular product is not necessary to the operation of an innovative environment. This can contain SMEs as well as large firms and even multinationals. In our view, the innovative environment is an "extended" district. In other words, SMEs and SMIs companies and multinationals can coexist there while benefiting from a certain degree of freedom; they maintain complementarities based both on specialized production of a product and on the need to achieve economies of scale.

4.4.3   The Role of Technopoles

These innovative environments are reminiscent of several examples that have characterized the economic vocabulary of recent decades. "Technopoles" and "technoparks" have become the two most popular types of industrial district among local authorities. According to Ruffieux, "technopoles or science parks are local geographic concentrations of innovative companies, located close to scientific research and training centres, with the aim of forming an innovative micro-system. (Ruffieux 1991: 375). This innovative micro-system, in turn, is characterized by fewer commercial relationships between companies, and by more specific research-industry relationships, and in particular relationships with local research. Ruffieux suggests that the birth of a technopole is characterized most of the time by spontaneous accidental phenomena, and at the same time by the slowness of the initial start-up process. However, in the case of Morocco, we do not believe in either spontaneity or accidental circumstances, as the region's social, legal and institutional environment does not allow for them.[16] To understand how these innovative environments evolved in different contexts, several relevant and innovative global experiences in the field of the technopoles are worth citing.

In Italy, an innovative governance approach to stimulate industrial innovation was launched in 2016, called the Industry 4.0 strategy. Despite criticism for its technological restrictions, the Italian government has created competency centres to facilitate technological transfer to businesses. The Emilie-Romagna region has played a leading role in developing a series of technoparks to promote business growth. Italian governance relies on a combination of top-down and bottom-up policies, involving various stakeholders in decision-making processes. The success of these technoparks, particularly in Emilie-Romagna, is attributed to effective synergies, international implications, and a solid industrial base.

The development of the French technopole of Sophia Antipolis was created through successive policies and public support. Despite the fact that this technopole successfully attracted technology firms due to strong external relationships, it suffers from a lack of creativity due to weak internal links between businesses. The collaboration among local actors is limited since many businesses are multinationals with headquarters outside of the region. Despite public sector involvement, the lack of internal synergies limits the technopole's innovative potential.

In Japan in 1980, the Ministry of International Trade and Industry (MITI) initiated a program to relocalize semiconductor production in Tokyo, Nagoya, and Osaka to address student unemployment. In 1983, the Japanese government passed a law to accelerate regional development around advanced industrial complexes, ordering 47 prefectures to plan local technopole creation programs. However, the success of these programs has been disproportional, with Japan hosting an average of 400 tech companies per technopole over the past decade. The main outcomes include the failure of the initial vision, the "branch plant" syndrome, the inability to develop links between universities and industry, lack of infrastructure, and inability to attract key workers.

In the United States, Silicon Valley had been agricultural land prior to the inauguration of Stanford Industrial Park in 1950. The creation of Frederick Terman, this can be considered the world's first tech park. Terman aimed to maximize innovation by transferring technology from East Coast centres, establishing networks, and collaborating with Silicon Valley transistors. Within 25 years Silicon Valley housed 90 businesses and 25,000 workers. This success can be attributed to strong government demand and space exploration.

---

[16] Several reforms are underway to adapt decentralization to a local economic approach. However, the territory is still not taken into consideration to increase non-market exchanges and cooperation between economic agents.

In the Moroccan context, the country does not have a base of innovative SMEs sufficiently large to allow such an approach. As a result, high-tech industry is synonymous with a renewal of the regional productive fabric and a new territorial dynamic. Although high-tech industries create few direct jobs, they nevertheless represent the pivot around which traditional industries will revolve. In this respect, Carroue (1989) asserts that "we shouldn't automatically expect new technological developments to have a direct and immediate impact on job creation and regional development."

For these reasons, we propose the adoption of "incubators"[17] associated with technology parks (Carroue 1989). We advocate for incubators mostly because they appear to be a new element in local authorities'interventions in regional planning and in the reshaping of their productive fabric. What is more, incubators do not align with the conventional concepts such as technology parks, science parks, or free zones that have been prevalent in the literature concerning the promotion of regional industrial development in Morocco. Although in our view it is the most appropriate innovative environment for such an objective, the incubator cannot systematically create a development dynamic. Hence, as a second step, we propose the utilization of "technology parks" to ensure technology transfer to modernize the national production-industrial fabric.

The incubator serves as a strategic and transitional choice. In other words, the incubator would be a new, preliminary stage in the creation of innovative and non-innovative SMEs. Once the incubator has reached maturity, the investment authorities will select, according to specific criteria, the innovative companies that qualify to be located in the technology park. The latter would thus be the ultimate phase of a possible complementarity between industry and research (Planque 1981).[18] Indeed, the technology park allows scientific research to be combined with technology. This is why it can provide a basis for regional development, thanks to the knock-on effects on the regional economy.

One of the key challenges for the incubator in our context is to increase the creation of SMIs and SMEs as part of this new industrial decentralization. However, business creation is being held back by the reluctance of owners of Moroccan capital. Transfixed by their investments in the tourism and real-estate sectors, they are reluctant to inject capital into high-tech industry. Consequently, the number of high-tech businesses being created is particularly low, and the ones that do operate belong to certain family-run companies and a few large national or foreign groups, Moreover, the rate of tech companies that close during the first 2 years of operating is very high. It would therefore require a great commitment from the Regional

---

[17] Faced with a profound crisis in planning and intervention tools for territorial change, this crisis calls for a search for ever more appropriate concepts and methods of intervention. The incubator is one response proposed by Carroue (1989). He defines it as an alternative to the painful regional effects of the redeployment of major industrial groups. The author sees it as a means of revitalizing innovative, job-creating SMEs.

[18] Planque demonstrates how new diachronic spatial dynamics unfold under the dual effect of major innovations in communication and information processing techniques.

Investment Centres to carry out "assistance" to regional, national and foreign businesses to minimize the risks of bankruptcy.

The stakes are high since high-tech industry is supposed to lead to a profound reorganization of the regional productive system and social changes, followed by a new urban hierarchy (Léo 1983). This potential urban dynamic presupposes the mixing of a certain number of socio-professional categories of diverse origins within the potential regions. Moreover, this dynamic will be based on the ability of regional cities to promote urban planning and assimilate business strategies in order to compete with other cities, particularly Casablanca. In fact, regional development is not solely linked to the "high-tech" industry, but also needs a land-use planning policy and financial decentralization to facilitate technology transfer, support for SMEs in the regional economy and to boost scientific research and development.

4.5 Scientific Research as a Lever for Regional Development

4.5.1 New Role of Universities

It is important to conceive of the university as a crucial actor and an undeniable reference in all activities and development strategies adopted by regional policies. However, in the Moroccan context, we must also act concretely on the deficiencies that burden the university system and impede, if not oppose, its adherence to the development dynamics initiated in various sectors. There is an urgent need to revive scientific research throughout the university, and promote scientific production. The ultimate objective is to build bridges between the university's laboratories and the local businesses within a region in order to build a Localized Production System (LPS).[19]

Technology Transfer involves the enrichment of research and development. It will enable the university to have a more significant presence in its region. In order implement such a project, we should bring together all the necessary ingredients, both traditional and conventional (considered as prerequisites) such as enhancing staff capacities at levels able to meet the required standards, implement appropriate and adapted teaching programs, recruit skilled people and provide enough resources to reach the objectives. Governmental institutions are engaged in this vision and strategy, as highlighted by Ahmed Réda Chami, President of the Economic, Social and Environmental Council:

---

[19] It is important to note that the industrial fabric of the Tangier-Tetouan-Al-Hoceima region is entirely delocalized and dependent on a number of foreign (mostly French) multinationals. It could be relocated at any time, depending on the international situation, depriving the region of considerable added value. The idea is to equip the region with a Localized Production System (LPS). This aligns with a long-term development vision, aiming to provide a sustainable guarantee for industrialists wishing to invest in the region.

> Universities are called upon to be drivers of regional development, to open up to the world, to strike a balance between theoretical academic programs and practical professional training, to be autonomous, and to adopt student-centered education to ensure their integration into the job market. These are the foundations of a university that will play a central role in the new model of development adopted by the kingdom." (le Matin, December 2021).

According to Chami, universities must become drivers of regional development by actively contributing to the balanced growth of various regions in the country, which would help reduce existing regional disparities. Moreover, to better prepare students for success in the job market, we have to maintain a balance between theoretical academic programs and practical professional training within universities. Additionally, the autonomy of universities is emphasized, allowing them to make strategic decisions in line with their missions while upholding educational quality standards. This discourse underscores that universities are key players in Morocco's New Development Model as engines of regional development, balancing their programs, enjoying autonomy, and focusing on student-centered education to facilitate their professional integration.

In the same context, Abdelatif Miraoui, Minister of Higher Education, Scientific Research, and Innovation emphasized that:

> The role of the university is also crucial for the country's development. It is even "at the heart" of all the transformations taking place in the world and must therefore contribute to the development of an individual's capacity to be autonomous and capable of transforming themselves to adapt to these changes and the significant technological advancements. . .".[20]

Miraoui focuses on the essential role of the university as a catalyst of economic development in Morocco. He emphasizes that the university should not only focus on academic education but also on preparing students to meet the evolving needs of the economy. This entails aligning academic programs with national priorities, promoting research and innovation, and strengthening the university's ties with industry. Accordingly, universities must strategize to achieve the economic development goals of the country.

Currently, the Moroccan university is perceived as a "black box" where the "output" is hardly in phase with the expectations that a university be more than just an academic institution, a place of knowledge acquisition and transfer. It should become a key player in developing conceptual intelligence and in becoming the incubator for innovation, which Morocco still sorely needs and sorely lacks. A qualitative leap in this direction is entirely possible. A holistic approach must be implemented where (a) basic principles will be to reformulate, (b) mode of management for scientific structures and activities will be more flexible and decentralized, (c) existing resources are optimized in a manner that respects the choices and orientations of researchers, and (d) training in and through research is prioritized over the medium and long term. In our view, this is how a university can serve as

---

[20] Le 360 (2022). Conference-debate on the theme 'The University as a Lever for Accelerating Socio-Economic Development in Morocco,' organized by the Faculty of Legal, Economic, and Social Sciences (FSJES) of Ain Chock, in Casablanca. https://youtu.be/UkJRoISuoYo

catalyst for in the field of scientific research and technology transfer, which has often been neglected in the past.

The effective involvement of researchers in the change is imperative in order to ensure its outcome and expected success. Such involvement should go beyond conducting sporadic consultations and periodic meetings among research players (i.e. the professors and researchers), to a genuine approach based on mutual trust and on continuous communication and exchange with all stakeholders.

4.5.2 Technology Transfer: A Way for Universities to Connect to Their Territorial Communities

Technology transfer has become a leitmotif concept in Moroccan political discourse. Directly related to the recovery and strengthening of the industrial sector, the State, particularly through the local authorities, tries, somehow, to create ideal and favourable conditions to boost regional development and promote an economy with strong added value. In Morocco, innovation was not a priority for the government until 2000, and there was a lack of connection between public research and the socio-economic sector. (Attou et al. 2019) The concept of incubation serves as an efficient method for linking technology, capital, and expertise to leverage entrepreneurial skills, expedite the growth of emerging enterprises, and consequently stimulate employment opportunities. Incubators are seen as promising tools for supporting innovation and entrepreneurial growth. Academic incubators focus on transferring scientific and technological knowledge from universities to businesses. (Attou et al. 2019) The introduction of incubation in universities through Law 01–00 revolutionized the higher education system and promoted the economic valorisation of research results.

> (...) within the framework of the missions assigned to them by the present law, universities may, by agreement, provide services for remuneration, create incubators for innovative companies, exploit patents and licenses and market the products of their activities..." (Royaume du Maroc 2000)

Given the context of Law 01–00, the need to put the university mission at the service of development and employment increased. This law has revolutionized the ways in which higher education and research operate. It has opened up new perspectives for universities by enabling them to participate in the creation of wealth of their territories and to contribute directly to regional and national dynamism. Thus, the long-term objective is to value research and place the university as a key technology producer, through cross-enrichment between academic research and industry. However, such an opening between the academia and the socio-professional world cannot be fully realized and successful if it does not also involve the region as a local authority in the desired transfer process. Of course, the latter depends first and foremost on industry, particularly high-tech industry. Nevertheless, it is also dependent on the region's spatial planning policy, which in one way or another enters into the technology transfer equation. Another equally favourable parameter can also contribute to this project. This lies in the evolution and implementation, on the ground, of advanced regionalization policy, which comes at just the right time. The latter provides greater autonomy and effectively makes the region more attractive and able to carry out its economic development.

With advanced regionalization, the region becomes a geographical, territorial, and administrative "ecosystem," entirely conducive to the success of technology transfer. With significant human and technical potential, some universities are also in the process of setting up "Innovation Cities," funded jointly by the Ministries of Higher Education and Scientific Research and Innovation, and the Ministry of Industry, Trade, Investment and the Digital Economy. Such an ecosystem will in itself contribute to exchanges between the University and the socio-professional environment as well as to the transfer of technologies. By creating an Innovation City, a university plans to become actively involved in the support and success of young innovative start-ups, in particular through the development of specialized services and the creation of various continuing education programs in technical, scientific and even environmental expertise for the benefit of operators.

Thus, for these universities, the creation of an "Innovation City" is an essential and decisive step towards realizing their commitment to the technology transfer process. However, no matter how effective such a structure may, it remains insufficient, on its own, to guarantee successful technology transfer. For such spaces to effectively serve as junction between research, development and training, and the business world, they need to host continuous intensive, collaborative interactions involving universities, researchers, industry representatives, business leaders, and government officials. If universities are to play an enhanced role in fostering innovation and regional development, three major actions are recommended:

1. First, reviewing the structure of scientific and technical research within universities so that they further exchange of good practices with the professional sphere,
2. Second, establishing a system of promotion and funding for excellence and high-value scientific technological production should be introduced,
3. Third, reinforcement of engineering training, and creating a new polytechnic structure within universities to broaden their operational technical base and to align it with local and regional specificities.

This reorganization of academic research will undoubtedly be fundamental in establishing the university as a key player in technology transfer and as a driver of regional development. However, for universities to assume this crucial role, the State must implement more effective strategies to facilitate collaboration between regional institutions and manufacturers.

4.6 Conclusion

Regional development in Morocco remains an ongoing process and more reforms are required to align decentralization with a localized economic approach. The new regional spatial organization, and Morocco's New Development Model might represent a fundamental shift beyond mere decentralization, offering opportunities for proactive regional initiatives. When it comes to actualizing technology transfer, incubators and technopoles could be a major model for local authorities. These hubs foster concentrations of innovative enterprises, closely situated near research and scientific training centres. Such spatial hubs create innovative microsystems where advanced industries operate. They serve as zones where university research and technology converge, able to catalyse regional industrial processes and, in turn, positively influence a region's development.

References

Attou OE, Taouaf I, Arouch M (2019) Les incubateurs universitaires au Maroc: État des lieux et perspectives. Int J Innov Appl Stud 26(1):163–174

Brousky L (1988) Essai sur l'aménagement du territoire au Maroc et ses implications institutionnelles et politiques. Dissertation,. Faculté de droit, Université de Lyon

Carroue L (1989) PME innovantes et développement régional: Les pépinières d'entreprises. In: Technologies nouvelles mutations industrielles et changements urbains, Collection Villes et territoires #2. Edition Presse Universitaire du Mirail, Toulouse

Courlet C, El Kadiri NA, Fejjal S, Ferguene A (eds) (2006) Territoire et développement économique au Maroc : Le cas des systèmes productifs localisés. L'Harmattan, Paris

Dagri T (2000) Collectivités territoriales et développement local : l'expérience marocaine. Faculté des sciences Économiques et sociales. Université des sciences et technologie de Lille, Lille

Dumas J (1989) Technologies nouvelles, développement régional et aménagement urbain : La politique française des technopoles. In: Technologies nouvelles mutations industrielles et changements urbains, Collection Villes et territoires #2. Edition Presse Universitaire du Mirail, Toulouse, pp 33–51

Farhat H (2004) Le développement décentralisé au Maroc, dynamique spatiale et planification régionale: le cas de la région de Chaouia-Ourdigha. Dissertation,. Université Michel De-Montaigne, Bordeaux 3

Friedmann J (2022) J. Friedmann on Agropolitan Development. In: The horizontal Metropolis: an anthology. Springer International Publishing, Cham, pp 481–485

Kahn R (1998) Facteurs de localisation, compétitivité et collectivités territoriales. Vues économiques 10:102–119

Léo P (1983) Les nouvelles formes de la mobilité des industries. In B Planque (eds), Développement décentralisé : dynamique spatiale de l'économie et planification régionale, Edition Litec, #16: 45-65

Marshall A (1898) Distribution and exchange. Econ J 8(29):37–59

Perrin J-C (1983) Contribution à une théorie de la planification décentralisée. In B Planque (eds), Développement décentralisé : dynamique spatiale de l'économie et planification régionale, Edition Litec, #16: 155–177

Peyrache V (1992) Le district industriel : un nouveau modèle d'organisation spatiale de la production et du développement régional. Problème économique 2(262):25–30

Planque B (1981) Innovation et dynamique spatiale des systèmes économiques. Dissertation, Faculté des sciences économiques, Université d'Aix en Provence

Rousset M, Basri D, Belhaj A, Garagnon J (1989) Droit administratif marocain. Imprimerie royale, Rabat

Royaume du Maroc, Ministère de l'Enseignement Supérieur, de la Recherche Scientifique et de la Formation des Cadres (2000) Loi n° 01–00 portant sur l'organisation de l'enseignement supérieur. Promulguée par le dahir n° 1.00.199 du 15 safar 1421 (19 mai 2000). Bulletin officiel n° 4798, Rabat

Ruffieux B (1991) Micro-système d'innovation et formes spatiales de développement industriel. In: Arena A (ed) Traité d'économie industrielle. Economica, Paris

Zaoual H (1996) Du rôle des croyances dans le développement économique. Dissertation, Faculté des sciences économiques, Université de Lille 1

Chapter 5
Financial Transparency and Sustainable Development in Local Governance in Francophone Sub-Saharan Africa: The Cases of Cameroon, Gabon, Burkina Faso and Côte d'Ivoire

André Akono Olinga

5.1 Introduction

Financial transparency and sustainable development are major issues in local governance (Manzanza Lumingu 2018). This is true of French-speaking Africa south of the Sahara in general, and of Cameroon, Gabon, Burkina Faso and Côte d'Ivoire in particular. They are consubstantial, for just as transparency is a component of sustainable development, the effectiveness of sustainable development depends on the transparent management of the funding allocated to it. Moreover, while it is true that sustainable development is a matter for multi-level governance (Minkonda 2018), it must nevertheless be recognized that local governance is nowadays considered the most appropriate stratum for conducting sustainable development projects. This was the rationale of a side event, called "Transparency, Integrity and the Fight Against Corruption: a key requirement for achieving sustainable development," of the United Nations Public Service Forum held Marrakech, Morocco, in 2018. The aim of this important event was to formalize the idea that there can be no sustainable development in Africa without transparency and integrity in local governance. The emphasis in this chapter is on financial transparency.

In the countries under study (Cameroon, Gabon, Burkina Faso and Côte d'Ivoire), public governance can neither ignore nor underplay the issue of sustainable

This chapter was translated from French by Ali Azeriah, Al Akhawayn University in Ifrane, Morocco.

A. Akono Olinga (✉)
Faculty of Legal and Political Sciences, University of Yaoundé 2, Soa, Cameroon

development in terms of social, economic and political contexts.[1] On analysis, governance has become one of the main dynamics of twenty-first-century public policy in the world in general, and across Africa in particular. This has happened despite the ambivalence of the stakes involved, which partially explains for the level reluctance and resistance to adherence to its philosophy. Sustainable development is no longer just a theoretical concept (Kabanda 2013). Today, it is the standard by which local public policies are conceived and implemented. It is on the governance agenda of most countries (Baudin 2009: 15). Sustainable development raises far-reaching cultural, political, social, economic, strategic and legal issue. The goal of sustainable development is peace and the present and future well-being of mankind. It is a development model with a human face that reconciles intergenerational interests. African local governance systems in general, and those of the countries under study in particular (Finken 2011), have explicitly or implicitly adopted this new paradigm. However, it should be noted that local finances, like state finances, are facing a resiliency crisis, and their management is exposed to the scourges of corruption and misappropriation. These dysfunctions logically limit their capacity to better provide for the financing of local development projects, hence the need for transparency in local financial management. In this respect, it seems that sustainable development in local governance is eminently conditioned by financial transparency.

Transparency is an ethical, political and scientific requirement (Bouvier 2010: 10). It takes on a specific dimension when transposed onto the public financial process. As such, it is a modern principle of public finance management, an innovation which stems from the new public financial governance (Ahouanka 2015). As a means of action, financial transparency is perceived as an institutional imperative and a management standard based on the accessibility, publicity, clarity and legality of budgetary and financial information during the budgetary process (Minkoueye Mi Nkoghe 2020). For all that, it is not just a procedural obligation (Guessele 2021), a determinant of good governance (Keutcha Tchapnga 2012), but the foundation of the new form of public financial governance (Sy 2017). financial transparency is a performance factor in the implementation of programs, including sustainable development projects and actions. A lack of transparency in public finance management leads to the evanescence of funds dedicated to financing sustainable development projects. The effective implementation of these projects, i.e., the satisfaction of the people, makes transparency a factor in strengthening not only local democracy, but also the legitimacy of local political leaders.

In any case, if the link between financial transparency and sustainable development seems obvious in local governance, we can still ask the question: does the practice of financial transparency contribute sufficiently to the achievement of sustainable development in local governance in the countries under study? Using a

---

[1] Beyond the concerns specific to each State, public governance in Africa is marked by the pursuit of the objectives of the African Union's Agenda 2063, whose aim is to catalyze sustainable development within the continent, and of the objectives of the United Nations' 2030 Agenda.

method that combines law and sociology, this analysis will first shed light on the reception of transparency and sustainable development in local governance. Then, it will clarify the fact that the contribution of financial transparency to sustainable development in local governance is still wanting. Finally, we propose ways of optimizing the contribution of financial transparency to sustainable development in French-speaking sub-Saharan African countries.

5.2   The Acceptance of Financial Transparency and Sustainable Development in Local Governance

Financial transparency and sustainable development might be recent developments, they are nonetheless already present in the local legal framework and are included in local development programs. Financial transparency has been accepted explicitly, unlike sustainable development, which has been accepted implicitly.

5.2.1   Local Governance's Explicit Commitment to Financial Transparency

The four local governance systems under study have not only adopted the democratic principles of public governance, but have also adhered to the philosophy of the new public financial governance. This philosophy is the foundation of performance and the driving force behind transparency in the management of local affairs. In Cameroon, Gabon, Burkina Faso and Côte d'Ivoire, transparency is formally enshrined and materially implemented in local governance. In essence, the principle of transparency is based on a requirement for truth in accounting and sincerity of figures. It obliges local authorities'accounting systems to comply with truthful information that respects accounting standards. This compliance is the guarantee of adequate and effective financing of sustainable development projects.

- In Cameroon, beyond Law #2018/012 of July 11, 2018 on the financial regime of the State and other public entities (Art. 4 Al. 9), local financial management is subject to the transparency requirement by Decree #2020/375 of July 07, 2020 on general public accounting regulations (Art. 1 Al. 2), and Law #2018/011 of July 11 on the Code of Transparency and Good Governance in the Management of Public Finances is quite clear (Art. 18) (Yatta 2014).
- In Gabon, financial transparency is regulated through Law #31/2010 of October 27, 2010 relating to finance laws and budget implementation, abrogated by organic law #020-2014 of May 21, 2015 relating to finance laws and law #021/2014 of January 30, 2015 relating to transparency and good governance in public finance management (Sorok and Bol 2021).

- In Burkina Faso, it is regulated through Law #008-2013/AN on the code of transparency in public finance management (Art. 1), Organic Law #73-2015/CNT of November 06, 2015 on finance laws (Art. 1), and Decree #2016-598/PRES/PM/MINEFID on general public accounting regulations.
- In Côte d'Ivoire there are the Organic laws #2014-336 of June 05, 2014 on finance laws, #2014-337 of June 05, 2014 on the code of transparency in public finance management (Art. 3), and Decree #2014-416 of July 09, 2014 on the general regulations of public accounting (Art. 1).

Thus, the linkage of local governance to the fundamental requirement of transparency is clearly set out in all four countries. In every case, local financial governance is at the heart of a vast reform movement that proceeds from the transformation of the local sector in a context of consolidating local democracy and boosting development. However, beyond the diversity of the standards mentioned, it is important to remember that they all share a common denominator: the sacralization of the requirement for financial transparency in local governance for effective sustainable development, albeit implicit in local normativity.

5.2.2 The Implicit Integration of Sustainable Development into Local Governance

While it is true that the integration of sustainable development into local governance is recent, it is no less true that its operationalization is still just emerging. This integration remains implicit, which can undermine the policy of operationalizing sustainable development in local governance in these four countries.

In essence, the implicit character is justified through the indirect consecration of sustainable development. In other words, the integration of sustainable development into the local normativity of the countries under study is deduced from certain provisions which seem, by interpretation, to express the realities of sustainable development. Moreover, in the absence of any direct or explicit enshrinement, it must be stressed that, given the subordination of local authorities to national legislation and in accordance with the principle of subsidiarity between the State and local authorities, sustainable development is well and truly present in local governance, despite the lack of clarity in its enshrinement.

By way of illustration, Articles 5, Paragraphs 2, 34, and 36 of Cameroonian Law #2019/024 of December 24, 2019 on the General Code of Decentralized Local Authorities, refer alternately to the promotion of development, the promotion of economic and social development, land-use planning, democracy and good governance at the local level (Donfack Sokeng 2007). This is also the case of Sect. 1.2, Chap. 1 of Title 2, which focuses on the environment and natural resource management without directly mentioning sustainable development. Moreover, the connection between local governance and sustainable development is more or less clear, albeit indirect, from an analysis of Article 39 Paragraph 1, which states that "local authorities shall carry out their missions in compliance with the Constitution and the laws and regulations in force". The laws and regulations in force include provisions on sustainable development, such as Law #2011/008 of May 6, 2011, which sets out the orientation for land-use planning and sustainable development, and Decree #2008/064 of February 4, 2004, which sets out management procedures for the national environment and sustainable development fund. In addition to these provisions, there are: Law #94-01 of January 20, 1994 governing forests, wildlife and fisheries, Law #96/12 of August 5, 1996 governing environmental management, and Law #2004-003 of April 21, 2004 governing urban planning.

In Gabon, Articles 2 and 219 of Law #001/2014 of June 15, 2015 relating to decentralization distinguish local authorities as players, in association with the national administration, in land-use planning, in economic, social, health, cultural and scientific development, and in environmental protection. On analysis, these are all areas that contribute to the achievement of sustainable development. It should also be remembered that local governance remains subject to State norms (Allogho and Nze 2013). This is the case also with Law #002/2014 of August 1, 2014 on the orientation of sustainable development.

The situation is similar in Burkina Faso and Côte d'Ivoire. In the former case, Articles 2 and 9, Paragraphs 1 and 4 refer to the promotion of local development in terms of economic, social, cultural and environmental development. This framework must also be in line with Law #2014-390 of June 20, 2014, the Sustainable Development Orientation Law. In the second case, Article 1 of Law #2003-208 of July 07, 2003 on the transfer and distribution of powers to local authorities likewise remains fixated on economic, social, health, educational and scientific development.

While there is a lack of clarity, i.e. an absence of the notion of sustainable development in these various texts, the fact remains that elements mentioned here and there within them are more or less related to sustainable development. What is more, local governance, in addition to being bound by the national legal framework, must also be in symbiosis with international, regional and sub-regional standards. This is the case with the African Charter of Values and Principles of Decentralization, Local Governance and Local Development adopted in 2014. It obliges local authorities to draw up their development plans incorporating the requirements of sustainable development (Art. 10, Al. 5). However, its implicit nature in local normativity may not only limit its operationalization, but also and above all, dilute the normative force of its legal framework at the local level. All of this means that financial transparency is not contributing sufficiently to the realization sustainable development.

5.3 The Incomplete Contribution of Financial Transparency to Sustainable Development in Local Governance

The contribution of financial transparency to sustainable development is real, but insufficient for several reasons. Broadly speaking, these include problems of governance, such as the proliferation of corrupt practices, and the gap between collective awareness and the challenges of sustainable development. In any case, it is important to identify the essence and meaning of the contribution before pointing out its shortcomings.

5.3.1 An Embryonic Contribution

The contribution of financial transparency to the achievement of sustainable development in local governance is more direct than indirect. From the point of view of direct contribution, transparency, both in the general and specific sense, has been identified as a sine qua non condition for achieving the UN's Sustainable Development Goals (SDGs). This is because financial transparency is a powerful force in the fight against corruption in the management and implementation of public policies, and therefore in sustainable development. It is even a requirement of sustainable development, which is linked to SDG 16 Peace, Justice and Effective Institutions. Clearly, we cannot have effective institutions without transparency in procedures, and even more so in the management of the funds allocated to them.

Indeed, transparency is consubstantial with sustainable development, hence its direct link with the latter. This close relationship can be seen in the theme of the aforementioned side event of the UN's Marrakech forum of 2018: "transparency, integrity and the fight against corruption: a key requirement for achieving sustainable development". This event offered the opportunity for participants to reaffirm the unwavering adherence of African states, and more specifically of local governments, to the standards of good governance, the fight against corruption, and the promotion of transparency and integrity, adopted by various African and international protocols. Among these are African Union's Agenda 2063 and its charters concerning the fight against corruption, as well as the strategies developed by the its member states to ensure that corrupt practices do not undermine the achievement of the 2030 SDGs (UCLG Africa 2018: 2). Indirectly, if this perspective is also to be retained, the quest for transparency contributes to the protection of funds dedicated to financing sustainable development. This in turn ensures that sustainable development projects are carried out.

Yet another aspect of contribution worth considering is fiscal transparency. According to the United Nations Conference on Trade and Development (UNCTAD), fiscal transparency through the fight against illicit financial flows is an effective means of contributing to the sustainable development of African countries. According to UNCTAD estimates, Africa loses around 88.6 billion US dollars a year to tax evasion. In this respect, it points out that curbing this phenomenon could make up half of the financing gap for Africa's SDGs, estimated at around 200 billion US dollars per year (UNCTAD 2020).

In any case, in Cameroon, Gabon, Burkina Faso and Côte d'Ivoire, the question of how financial transparency can contribute to sustainable development in local governance no longer arises. In addition to being a criterion in its own right, financial transparency contributes directly and indirectly to sustainable development. The focus now should rather be on the quality of this contribution. In this respect, the contribution is incomplete, given the limitations it faces in a context of underdevelopment.

5.3.2   A Failing Contribution

The failure of financial transparency to contribute to sustainable development in the countries under review stems from both structural and cyclical limitations.

On a structural level, the lack of financial transparency is linked to the fragility of the institutional framework in charge of controlling the management of public finances in general, and of local finances in particular. To be sure, there can be no financial transparency without real control, which means dynamic control (Akono Olinga 2020). To put it plainly, dynamic control is fundamentally variable, even flexible. It is a form of control that evolves over time, and integrates the societal realities of the geographical area in which it takes place.

The institutional framework in charge of controlling the management of local finances is diversified, but relies essentially not only on the contribution of local deliberative bodies, but also and above all on the action of the courts of audit (Begni 2017; Djeya Kamdom 2020). The judge is the guarantor of the legal order and the condition for the realization of rule of law (Chevallier 1994: 13). However, analysis of the operational contribution of these bodies to the achievement of financial transparency at local level reveals, firstly, deliberative bodies whose actions are politicized and further weakened by asymptotic decentralization. Secondly, regardless of their various designations (*Cour des comptes, juridiction des comptes* or *juridiction financière*), the failure of the audit jurisdictions of the countries under study can still be traced to the problem of their independence (Ouedraogo 2013).

At an economic level, the countries covered by this study are among those whose governance is plagued by the scourge of corruption. To put it plainly, today more than ever, corruption has an economic cost for local governance which is not without consequence for local development efforts in general, and the need to satisfy local needs in particular (Bouvier 2010; Cartier-Bresson 2000). As a result, it contributes to the evanescence or evaporation of public finances, which are essential for financing local public policies (Bensouda 2007). Perceived differently, corruption is not only a moral problem, it also reveals a political dysfunction with deleterious economic consequences (Chevauchez 2000), thus contributing to the perversion of the normal functioning of democracy. Local financial governance is not spared this scourge, which amplifies as it takes new forms in a context of crisis. The public finance crisis is partly to blame for the rise of corruption in public governance systems in the twenty-first century. In this context, it is difficult to set up a framework for implementing and expressing transparency. In the four States under review, financial transparency appears more as a political slogan than an administrative practice.

In any case, the lack of transparency in the management of local finances leads to the diversion of these finances and consequently limits the financing and implementation of sustainable development projects. It is beyond urgent for the States in question to implement adequate measures to optimize the contribution of financial transparency in local governance to sustainable development. For there can be no democracy without clear public finances, transparency and the rule of law (Gaba 2000). Opacity in matters of public finance is obviously the source of all abuses.

5.4 Optimizing the Contribution of Financial Transparency to Sustainable Development in Local Governance

Financial transparency is a cardinal principle of public finance management and local governance as it is a prerequisite for the proper implementation of local public policies. At the same time, it is a prerequisite for the realization of democracy at the local scale. To make the best possible contribution to the achievement of sustainable development, financial transparency must transcend the formal framework and become more real. This could be achieved by strengthening the legal foundations of sustainable development at the local level, and by optimizing the levers or dynamics of transparency.

5.4.1 Strengthening the Legal Foundations of Sustainable Development in Local Governance

The contribution of financial transparency to the achievement of sustainable development cannot be effective if sustainable development itself does not have a clear legal basis capable of reinforcing its normative force. In the same vein, it should be remembered that the achievement of sustainable development through the attainment of its objectives in local governance is conditioned by the normative force of its legal framework. The local normative framework is merely an extension of the national legal framework, which itself is not sufficiently dense and explicit in enshrining sustainable development. The constitutional norms of the countries under study are a perfect illustration of this. In French-speaking sub-Saharan Africa in general, sustainable development is implicitly enshrined in the constitution, hence the urgent need to explicitly enshrine sustainable development in the spheres of national governance in general, and local governance in particular.

Briefly put, the inclusion of sustainable development in the constitutions—as well as the texts which structure decentralization—of French-speaking sub-Saharan African states will reinforce the interest already accorded it in governance policies and programs. In this way, it will be clearly elevated to the level of highest national interest. It will also help to reinforce the mandatory and binding nature of its legal framework at local scale. What is more, a legal framework can really help achieve the objectives for which it was designed only if its binding and obligatory nature is effective and, above all, asserted. This effectiveness implies the conformity and regularity of the development policies pursued by local authorities. In principle, legal rules are binding and obligatory, compelling those to whom they are addressed to comply with their prescriptions. On analysis, this principle is effective in the case of "hard" law. However, in the case of sustainable development, its legal framework in local governance would fall under "soft" law, given the embryonic nature of its rules.

Consolidating the constitutional foundations of sustainable development also means affirming the value of the SDGs in local legislation. This should be achieved by highlighting them in local texts. This will reinforce their value and ensure that they are taken into account in local development programs. As a public governance objective, the SDGs are a tool for assessing sustainable development, which in turn is the perfect translation of the UN's 2030 Agenda for Sustainable Development. In essence, the said program, which is considered the blueprint for achieving sustainable development, was officially adopted in September 2015 in New York. It is the action plan for people, for the planet, for prosperity and for peace. It fosters a vision for the transformation of the world through eradication of poverty, and for its transition to sustainable development.

The strengthening of the legal framework for sustainable development will make the validation of local development programs subject to compliance with sustainable development requirements. That said, financial transparency alone is not enough to achieve sustainable development in local governance. It must be accompanied by a normative and institutional structuring of sustainable development at the local level to produce the desired effects, even if it must also be recognized that the consolidation of the vectors of financial transparency remains indispensable.

5.4.2 Consolidating the Vectors of Financial Transparency at the Local Level

Achieving sustainable development in local authorities in Cameroon, Gabon, Burkina Faso and Côte d'Ivoire also depends to a great extent on consolidating the vectors of financial transparency at the local level. This is crucial to the success of sustainable development projects. To achieve this, we believe two main measures are necessary. These are: boosting political control of local finances, and increasing citizen control of local governance in general.

In essence, while political control of local finances is in place in these countries, it lacks vitality for reasons inherent in the decentralization process. In a context of territorial decentralization (Nach Mback 2003), control over the management of local finances is primarily, but not exclusively, the responsibility of the deliberative bodies of local authorities. Political control by the deliberative assemblies comes under what is known as a posteriori internal control (Levoyer 2007). This involves ascertaining that budget execution operations have been carried out in accordance with the authorizations granted. However, this type of control, although more legitimate, is hampered not only by political considerations, but also by administrative ones, such as the thorny problem of the hypertrophy of supervision over local authorities.

With regard to the expansion of citizen control, as a corrective to the action of deliberative bodies, this will undoubtedly contribute to optimizing transparent management. After all, citizens are the main beneficiaries of projects designed to meet local needs. Citizen control in local governance in Africa is emerging (Ngueche 2018). The Institut de Recherche et de Débat sur la Gouvernance (IRG) defines citizen control as: "all practices, whether collective or not, sectoral or general, aimed at ensuring the accountability of players involved in the management of public affairs, notably through greater transparency" (Samb 2014). In short, it is a form of control that strengthens citizen participation in the management of public affairs (Pekassa Ndam 2014). There can be no citizenship without participation in public life (Njoya 2016).

Citizen control is an instrument for expressing and realizing the right to information in the management of public and therefore local affairs. It is a fundamental right that belongs to the citizen (Nkouayep 2018). Moreover, it has been enshrined in international legal instruments adopted by all the States under study. These include the Declaration of Human Rights and of the Citizen, which states that "society has the right to call to account any public official of its administration" (Déclaration des Droits Humains et des Citoyens 1789: Art. 15). This implicit reference reflects the fact that society needs to be informed. The African Charter on Human Rights and Peoples' Rights is more explicit. It is interesting in that it is the only text to refer directly to the right to information. In essence, it states that "everyone has the right to information" (ACHRPR 1981: Art. 9).

In short, political control and citizen control are instruments for consolidating transparency in the age of performance and the quest for sustainable development. Their adaptation to the local context is necessary for local governance to promote sustainable development in Africa.

5.5 Conclusion

Financial transparency is a requisite of sustainable development and a prerequisite for achieving the SDGs. There are two main reasons for this. First, it is a means of building a state based on the rule of law, and no development, however sustainable, can be conceived without the rule of law. Secondly, sustainable development is a set of programs, projects, activities and actions that cannot be carried out in the absence of funding, the availability of which can only be guaranteed by transparency. In addition, other measures complementary to those mentioned in this analysis are essential. According to the Declaration adopted by participants at the Marrakech event in 2018, local authorities must:

- Make a firm commitment to promote transparency and integrity in the governance of towns and territories by adopting citizen service commitment charters;
- Work to consolidate participatory democracy by putting in place innovative approaches to involve the population in defining and implementing local public action priorities, such as participatory budgeting; and
- Promote and develop an observation and monitoring system for the follow-up and evaluation of local public policies.

In addition, for these measures to be effective rather than aspirational, States must pursue efforts to strengthen administrative and financial autonomy (Essono Ovono 2009). Achieving sustainable development through financial transparency also requires this. The momentum is there, it's just a matter of time.

References

African Charter of Human and Peoples' Rights (1981) Adopted June 27, 1981, OAU Doc. CAB/LEG/67/3 rev. 5, 21 I.L.M. 58 (1982)

Ahouanka ES (2015) La transparence budgétaire dans les Etats membres de l'Union Economique et Monétaire ouest-africaine : Réflexion sur le nouveau rôle du parlement et de l'opinion publique. In: La LOLF dans tous ses états. Centre des Publications Universitaires, Abomey-Calavy, pp 535–563

Akono Olinga A (2020) L'apport de la performance au contrôle des finances locales au Cameroun. Dissertation, Université de Yaoundé 2

Allogho NF, Nze BL (2013) S'inspirer et se comparer : Institutions locales et gouvernance de la décentralisation au Gabon. In: Ango Ela P (ed) Les politiques de la décentralisation au Cameroun: Jeux, enjeux et perspectives. l'Harmattan-Cameroun, Yaoundé

Baudin (2009) Le développement durable, nouvelle idéologie du XXIe siècle. L'Harmattan, Paris

Begni B (2017) Le principe de transparence dans les finances publiques des Etats membres de la CEMAC. Revue Africaine des Finances Publiques 2:199–230

Bensouda N (2007) Efficacité et transparence des finances publiques pour une meilleure offre de biens pour le citoyen. Revue Française des Finances Publiques 100:333–336

Bouvier M (2010) Les transformations de la légitimité de la gouvernance financière publique. In: Mélanges en l'honneur de Pierre Beltrame. P.U.A.M., Marseille

Cartier-Bresson J (2000) L'analyse des coûts économiques de la corruption. Revue Française de Finances Publiques 69:19–32

Chevallier J (1994) L'Etat de droit, 2nd edn. Montchrestien, Paris

Chevauchez B (2000) Corruption et gestion publique. Revue Française de Finances Publiques 69: 87–94

Djeya Kamdom YG (2020) La réforme du contentieux financier public au Cameroun par la loi du 11 juillet 2018: portée et insuffisances d'un texte. Gestion & Finances Publiques 6:123–128

Donfack Sokeng L (2007) Bonne gouvernance, Etat de droit et développement: Approche critique de la réforme de l'Etat en Afrique. Revue trimestrielle de droit et des activités économiques (2), Presses Universitaires Libre 241–296

Essono Ovono A (2009) L'autonomie financière des collectivités locales en Afrique noire franco-phone: Le cas du Cameroun, de la Côte-d'ivoire, du Gabon et du Sénégal. Afrilex (special issue Finances Publiques, 2nd ed)

Finken M (2011) Gouvernance communale en Afrique et au Cameroun. l'Harmattan, Paris

Gaba L (2000) L'État de droit, la démocratie et le développement économique en Afrique subsaharienne. l'Harmattan, Paris

Guessele LP (2021) L'exigence de transparence dans les finances publiques des Etats de la CEMAC. In: Mede N, Debene M (eds) Les finances publiques entre globalisation et dynamiques locales, Mélanges en l'honneur de Eloi Diarra et Salifou Yonaba. Dakar, L'Harmattan-Sénégal

Kabanda K (2013) Le développement durable en droit congolais de l'environnement: engagements et limites face aux enjeux d'un droit en construction. Revue Africaine du Droit de l'Environnement 00:123–128

Keutcha Tchapnga C (2012) L'influence de la bonne gouvernance sur la relance de la décentrali-sation territoriale en Afrique au sud du Sahara. Revue Africaine de Droit Public 1(1):151–166

Levoyer L (2007) Finances locales. Hachette, Paris

Manzanza Lumingu MM (2018) Droit, bonne gouvernance et développement durable. In: Manzanza Lumingu MM (ed) Méllanges en l'honneur du Professeur Jean-Michel Kumbu Ki Ngimbi. L'Harmattan, Paris

Minkonda H (2018) L'émergence de la gouvernance multi-niveaux dans la communauté économique et monétaire de l'Afrique centrale à partir du programme économique régional: le cas du Cameroun. Dissertation, Université de Yaoundé 2

Minkoueye Mi Nkoghe E (2020) Réforme budgétaire et modernisation de la gestion publique au Gabon. Dissertation, Université de Montpellier

Nach Mback C (2003) Démocratisation et décentralisation: Genèse et dynamique comparées des processus de décentralisation en Afrique Subsaharienne. Karthala-PDM, Paris

Ngueche S (2018) Le contrôle par le citoyen des finances publiques au Cameroun. Revue Africaine de Droit Public 7(13):155–183

Njoya J (2016) Gouvernance locale et démocratie 'représentative' au prisme de la participation citoyenne. Revue Africaine d'Études Politiques et Stratégiques 1:7–20

Nkouayep LCP (2018) Le droit à l'information du citoyen local en droit public financier camerounais. Revue Africaine des Finances Publiques 3–4:11–42

Ouedraogo D (2013) L'autonomisation des juridictions financières dans l'espace UEMOA. Étude sur l'évolution des Cours des comptes. Disseratation, Université Bordeaux IV—Montesquieu

Pekassa Ndam G (2014) La participation avec gestion de budget: concept et enjeux d'une gouvernance territoriale en Afrique noire francophone. In: Guglielmi G, Zoller E (eds) Trans-parence, démocratie et gouvernance citoyenne. Panthéon–Assas, Paris/Lextenso, pp 199–218

Samb NM (2014) Gouvernance territoriale et participation citoyenne au Sénégal. Dissertation, Université de Montpellier III—Paul-Valery

Sorok A, Bol PG (2021) Constitution et bonne gouvernance dans les Etats d'Afrique noire francophone. Revue Africaine de Droit Public 10(24 supplément):313–342

Sy A (2017) La transparence dans le droit budgétaire de l'Etat en France. L.G.D.J, Paris

United Cities and Local Governments of Africa (2018) the UCLG Africa Annual Conference in Marrakech, 2018. https://www.uclga.org/news/fighting-corruption-and-promoting-integrity-and-transparency-at-the-local-level-in-africa-time-to-act-is-now/

United Nations Conference on Trade and Development (UNCTAD) (2020) Economic Develop-ment in Africa Report

Yatta FP (2014) La gestion des finances locales en Afrique: Convergence et Divergence des systèmes. Revue Africaine des Finances Locales: 5–34. Accessible on https://www.uclg.org/sites/default/files/revue_africaine_des_finances_locales_1.pdf

Chapter 6
The Developmental Role of Local Authorities: The Case of Tunisian Communes

Mohamed El-Mensi

6.1 Historical Overview

The municipal institution is well anchored in Tunisia's political-administrative landscape. The creation of the first municipality (Tunis) dates to 1858, a time when the country's territorial organization was essentially built around tribal lines, allowing sub-national entities (few communes, and the *caïdats*)[1] to enjoy relative autonomy in the running of day-to-day local affairs, despite the fact that their respective territorial boundaries were not clearly demarcated. After 1881 (when the French protectorate was established), the colonial authorities sought to "modernize" the administrative system in order to tighten control over territories and populations. The policy followed was to weaken traditional ties by strengthening the administrative model. A new territorial configuration was put in place, headed by colonial administrators (*contrôleurs civils* in French), for more systematic control of territory and tribes. At the same time, the municipal system gradually expanded from 1884 onwards (creation of six new communes), to grow in number throughout the rest of the colonial era, mainly in urban areas populated by Europeans. The unelected municipal councils were dominated by French notables.

After independence (1956), the national authorities pursued the policy of breaking with the traditional (tribal) system of territorial and administrative structure

---

[1] The *caïdat* is a regional office of the beylical administration. The *caïd* had all the prerogatives of a governor of a province. He was responsible for general administration, the maintenance of justice and the lucrative position of tax farmer. He was the enforcement agent of the Bey of Tunis, and presided over the destiny of the region, its tribes and surrounding villages. The caïd-governors are part of the beylical makhzen (political-administrative body of the state).

M. El-Mensi (✉)
Governance and Local Development Consultant, Senior partner - Local Development International, New York City, NY, USA

© The Author(s), under exclusive license to Springer Nature Switzerland AG 2024   75
K. Darmame, E. Ross (eds.), *Local Governance and Development in Africa and the Middle East*, Local and Urban Governance,
https://doi.org/10.1007/978-3-031-60657-1_6

initiated under the French protectorate, the main objective of which was to reinforce the authority of the new State and the prerogatives of the central administration. In this respect, the ideas of a *strong State* and *stable political regime* were recurrent themes in the debates of the National Constituent Assembly (L'Assemblée Nationale Constituante, 1957–59), which explains the choice made at the time to perpetuate the same system of administrative organization inherited from the colonial period. This was a hyper-centralized organization, with political and economic power monopolized by the central authority, aided by its "extensions" in the territories, represented by the institution of the Governor (and his deputies, the *délégués*) and the deconcentrated branches of the sectoral ministries.

Indeed, the first constitution of independent Tunisia (adopted in 1959) made no explicit reference to the principle of decentralization. It used a lapidary formulation—an "orphan" provision—stating that "municipal councils, regional councils and structures to which the law confers the status of local authority manage local affairs under the conditions laid down by the law." The 1959 Constitution remained therefore silent on the principles that could have underpinned true decentralization, such as the elective nature of deliberative bodies and management autonomy. The framers of the Constitution had therefore opted for a rather juridistic-administrative approach to the status and prerogatives of local authorities. Thus, the organization of the national territory was modified on several occasions, mainly in response to security and control imperatives, and sometimes for technical or developmental purposes. The number of municipalities rose from 69 in 1956 to 134 in 1966, eventually reaching 350 in 2016, while the national territory was divided into a number of governorates (*wilaya*), which were initially deconcentrated administrative districts, before subsequently being transformed into hybrid structures (with a "zest" of devolution) by the creation of unelected Regional Councils, chaired by the governors (1989).

6.2 New Order: The 2014 Constitution

The 2014 constitution marks a break with the old order of administrative and territorial organization, in response, it seems, to the demand for 'territorial' equity forcefully expressed during the events of December 2010–January 2011.[2] It affirmed decentralization as a principle of territorial organization and a path towards the mitigation of development disparities between the country's different regions. The 2014 Constitution [Art. 12 and 14] commits the State "to strengthening decentralization and applying it throughout the national territory," while ensuring "a balance between regions by reference to development indicators and based on the principle of positive discrimination."

---

[2]Referred to usually as the start of the so-called 'Arab Spring'.

The decentralization model decreed by Art. 14 of the 2014 Constitution provides for a territorial organization structured into three tiers of subnational authorities. To the existing communes, are added regions and districts (a supra-regional tier), covering the entire national territory.[3] All three levels of local authority have legal personality and financial and administrative autonomy, and they are supposed manage their affairs in accordance with the principle of self-administration, under the authority of elected bodies. They are assigned own functions, functions shared with the central government, as well as any additional functions that the central government could transfer, in the future, in accordance with the principle of subsidiarity. The Constitution also commits the State to setting up a fiscal transfer mechanism, so that any creation or transfer of local functions must necessarily be accompanied by the assignment of corresponding resources. The institutional structure thus created is headed by the Higher Council for Local Authorities (*Conseil Supérieur des Collectivités Locales*), a representative body of subnational authorities tasked, among other things, with addressing issues relating to local development and regional disparities. The constitutional provisions were subsequently spelled out in legislative rules, the Local Authorities Code (Code des Collectivités Locales, CCL).

The decentralization reform was ambitious, and the expectations of citizens, particularly those in the interior regions of the country, were high, at least at the outset. They had high hopes that the existing territorial "divide" in terms of development would soon be reduced, if not bridged, thanks to the action of local authorities, autonomous and equipped for the task, and to the promises of the "positive discrimination" clause enshrined in the Constitution. In other words, putting in practice the Constitutional principals and the provisions of the CCL is a long road that requires determination and commitment on the part of the national authorities, and support and participation on the part of everybody. The challenges to be met are enormous, and the obstacles to the emergence of genuine developmental local authorities are numerous and significant.

6.3 Autonomy and Status to Be Asserted

The first challenges are the affirmation of local autonomy, i.e. how to put local authorities (LAs) in a position to effectively fulfill their role as actors of local development, and the mitigation of regional disparities. Admittedly, the Constitution enshrines "autonomous (self) administration"[4] as a basic principle of local action, but the scope of autonomy and the ways in which it is practiced were left to the law

---

[3]Previously, communes covered only urban areas. The number of communes increased from 264 to 350, with the creation of 86 new communes and the extension of the territorial limits of the remaining communes to include adjacent rural areas.

[4]In application of this principle, local authorities are, in principle, subject to ex-post control of the legality of their acts, which is exercised under the supervision of the competent financial and administrative courts.

to define. This is how the *Code des Collectivités Locales* (Local Authorities Code) reaffirms this principle, notably by ruling out any ex-ante control by the central authority over local decisions, and by granting local authorities, for example, the discretion to decide on the pricing of paid services (taxes and duties, however, remain the jurisdiction of the State). However, there are limits to this autonomy. Indeed, the Governor (the State's representative at the territorial level) has the power to challenge local authorities' decisions in several areas of governance, such as: awarding public contracts, subsidies to the local private sector, rates for local services, voting of the local budget, modification of budget appropriations, balancing of budget execution, regulatory decisions, etc. This intrusion of the central authority into the local sphere is counterbalanced, however, by the Local Authorities' right to bring the matter before the administrative courts, in accordance with a procedure that could prove cumbersome (with appeal, if necessary, to the judgments handed down in the first instance). There is therefore a risk that regional administrative courts will be clogged with a multitude of appeals, for which they are not equipped to deal with in a timely manner.

Furthermore, a change in the political situation and in the constitutional order took place in 2021,[5] resulting in a new constitution adopted by referendum (July 25, 2022), and in new government practices that raise doubts about the new regime's intentions regarding the status and prerogatives of local authorities in general, and of municipalities in particular. Indeed, the 2022 Constitution takes a step backwards from the 2014 Constitution in terms of decentralization, reverting to a narrow, minimalist conception of the role of local authorities, identical to that of the 1959 Constitution.[6] It is still too early to say with any certainty whether this is the emergence of a new paradigm, likely to call into question the recent gains in decentralization by reactivating entrenched centralizing practices (subordination of local authorities to the central government). Since July 2021, there are signs and practices that tend to justify this fear. These include: the abolition of the Ministry of Local Affairs, the renewed subordination of local authorities to the Ministry of Interior, and recurrent interference by governors in the municipal affairs (suspension of decisions, obligation to go through the governor to communicate with the central authorities). Most worrying of all, however, is the uncertainty surrounding the future status and role of local authorities, and their relationship with the two houses of Parliament, the People's Assembly, and the National Council of Regions and Territories, foreseen by the 2022 constitution. Indeed, the electoral system for both houses is that of a single-member constituency system at the level of the Delegations (territorial subdivisions of the governorates), which de facto excludes political parties. Such an electoral system, with its associated territorial division, could lead

---

[5] Dismissal of the government and suspension, then dissolution of the Parliament by the President of the Republic, on the pretext of imminent danger threatening the functioning of State institutions (application of Article 80 of the 2014 Constitution).

[6] Article 133 of the 2022 Constitution mirrors in the same terms Article 71 of the 1959 Constitution, thus wiping the slate clean of the previous detailed provisions, contained in the 2014 Constitution.

to confusion, or even to competition between members of the parliamentary houses and local elected representatives for the same electoral base and for the provision of services to the population, to the detriment of the proper functioning of local authorities. This is particularly obvious in the case of the regions, which may be perceived less as a territorial authority and more as a springboard for access to the membership of the National Council of Regions and Territories.

6.4 Competences to Be Clarified and Consolidated

As the regions and districts have not yet been established, only the municiplaities are currently operational following the local elections held in May 2018 in the wake of the promulgation of the Local Authorities Code (CCL).[7]

Municipal functions are of three types[8]: (i) own competences centered, for the most part, on the provision of local public services and infrastructure; (ii) joint competences (shared with the central administration) which relate, in particular, to the realization and operation of sectoral services and facilities (health, transport, tourism, environment, social welfare, culture, roads, education), and (iii) transferable competences—those that the central administration could later relinquish to the municipalities if necessary, such as: construction and maintenance of health facilities, educational institutions, and cultural and sports facilities.

In practice, municipalities are currently confined to providing standard proximity services; they are not involved in the development or operation of sectoral facilities and services which are the responsibility of the relevant ministries and State-owned enterprises. Nor is there any indication that additional powers will be transferred to the municipalities any time soon. And yet, the Local Authorities Code [Art. 200 and 238] assigns to municipalities a general mandate to work for the "economic, social, cultural, environmental and urban development" of their territory, and to "take the necessary measures to promote the development of the municipality, and increase attractiveness for investment through the realization of infrastructures and public amenities."

Thus, the functional autonomy of local authorities (scope of competences) is limited and unclear. The CCL and its implementing regulations published to date do not provide the necessary clarification of local competences and the conditions under which they are exercised, and more particularly of the rules for exercising the *general power of competence* assigned to the municiplaities. The latter's own functions are unchanged compared to the past (mainly proximity services), while

---

[7] At the regional (governorate) level, the current regional councils continue to function as local authorities, under legal provisions dating from 1989, pending the organization of the first regional elections (the date of which has not yet been set) and the establishment of the regions as a second tier of local authorities—to replace the current regional councils.

[8] In addition to the general power of competence clause.

those of the future regions are limited. Also, the areas of competence transfer are addressed in the CCL in very general (and sometimes vague) terms, and are limited in scope for the time being. The same applies to powers shared with the central authority, the exercise of which will require cooperation between local authorities and deconcentrated State administrations. However, it is not yet clear whether this cooperation is to be established on a statutory basis, or whether it will result from agreements reached on a case-by-case basis.

Ultimately, the degree of local functional autonomy (and, consequently, its impact on local development) will depend on the scope of local competences, i.e. the choice of to be made by the national authorities (government and parliament) between two attitudes: (i) granting local authorities the competences required (with adequate funding sources) to enable them to fully assume their developmental role, or (ii) adopting a restrictive approach by limiting the scope of local competences (which is already reflected in the Code), arguing for lack of local financial and institutional capacities, for example.[9] Another pitfall to be feared is that of central authorities "offloading" certain tasks to local authorities, without the latter having the means to carry them out in a uniform and equitable manner throughout the national territory, which would end up further exacerbating regional disparities.

6.5   Structural Financial Imbalance to Be Corrected

Municipalities currently finance their activities through their own-source revenues and fiscal transfers, in addition to the possibility of borrowing from a State institution, the Loan and Support Fund for Local Authorities (*Caisse de Prêts et de Soutien des Collectivités Locales*, CPSCL).

Own-source revenues are made up of local taxes,[10] fees and charges for the use of local public services,[11] and revenues generated by the operation of other elements of the municipal assets. Taxes and similar levies that municipalities are entitled to collect are established by law; tax bases are determined, depending on the case, either by the municipalities themselves (property taxes) or by the State tax authorities (taxes on economic activities), while the rates are set by law, and collection is the responsibility of the Ministry of Finance. On the other hand, non-tax revenues (notably market concessions and similar revenues) are assessed and collected by the municipal staff and/or the municipal accountant (*receveur municipal*).[12] It should

---

[9] A decisive decentralization implies the transfer of coherent, complete blocks of competencies. A partial transfer that does not achieve full decision-making autonomy, because the State has retained one part of the decision that conditions the rest, would be contrary to the spirit of decentralization.

[10] In particular: tax on industrial, commercial or professional establishments, property tax (on buildings and vacant urban land); hotel tax.

[11] In particular, revenues from concessions to operate municipal markets.

[12] Assigned to the municipalities by the Ministry of Finance (Treasury Department)

be noted, by the way, that the discretionary taxing power of municipalities is extremely limited, insofar as the State has almost total control over the tax chain (creation of taxes and levies, determination of bases and rates, collection). Similarly, the lack of resources and/or motivation of the Ministry of Finance's staff means that the yield from local taxation is not optimized.

Fiscal transfers are intended to make up for the shortfall in own-source revenues; they consist of operating and capital grants:

- The operating grants come from the Decentralization, Equalization and Solidarity Support Funds (*Fonds d'Appui à la Décentralisation, de Péréquation et de Solidarité entre les Collectivités Locales*).[13] This Fund is currently financed by an annual contribution from the national budget[14] which is distributed among the municipalities based on five criteria.[15]
- Capital grants are managed by the Loan and Support Fund for Local Authorities (CSPCL) on behalf of the State.[16] They are of two types: (a) performance-based general-purpose (non-earmarked) grants, awarded to all municipalities, half of which according to three criteria (population, fiscal potential, and underdevelopment index), the rest according to performance score,[17] and (b) purpose-specific (earmarked) grants intended to finance projects in disadvantaged municipal areas.

The operating grant is permanent, insofar as it is based on a legislative provision (the aforementioned Decentralization, Equalization and Solidarity Support Funds), whereas the capital grant mechanism has been financed until now through a national program,[18] implemented with the financial assistance of the World Bank (budget support loan), which is due to expire at the end of 2023. There is therefore some uncertainty as to the continuity of this mechanism beyond this date.

Review of municipal finances reveals a chronic self-financing deficit[19] (insufficient own-resources), which is partially made up by fiscal transfers, reflecting a decline in the municipal financial autonomy index[20] (53.4% in 2020 according to the

---

[13] Established by Law no. 2020-46 of December 23, 2020, the Finance Act for the year 2021.

[14] The law also provides for other sources of financing for the Fund, allocation of a share of the proceeds from certain taxes, and where appropriate, a share of State revenues from the exploitation of natural resources.

[15] Equal shares (10%); population (40%); three-year average of revenues collected from property tax (31%); population of municipalities whose own-source revenues are below the national average (9%); the remainder (10%) is allocated to municipalities faced with financial difficulty.

[16] As part of the Urban Development and Local Governance Program (UDLGP), financed in part by the World Bank.

[17] Performance is measured annually in three areas: Service improvement, Participation and transparency, and Resource improvement.

[18] Urban Development and Local Governance Program (PDUGL).

[19] This deficit has been exacerbated by the inclusion, in municipal territories, of rural areas inadequately endowed with local services and infrastructure, and poor potential for raising own revenue.

[20] Financial autonomy is measured by the share of a municipality own-source revenues in its total available financial resources.

latest figures available),[21] meaning heavy dependence on the State aid. Another indicator of the municipalities' financial "fragility" is their endemic level of indebtedness, which continues to worsen despite recurrent intervention by the State through rescheduling and restructuring measures. At the end of 2020, outstanding municipal debts stood at 191.7 million dinars, or 29% of their own-source revenues. However, this average masks disparities between individual municipalities (for some of them, outstanding debts equal more than 100% of their own-source revenues).[22] The main creditors of the municipalities are State institutions (electricity distribution company, and CPSCL).

6.6 Deconcentration: To Be Furthered

Reforming the system of territorial and administrative deconcentration to align it with the decentralization policy is another challenge that needs to be addressed. As mentioned above, public administration in Tunisia still bears the marks of a centralizing tradition that goes back a long way in the country's history. Indeed, territorial control by the Center has been a constant feature of the administrative practices, both before and during the colonial period, and since the independence. The institutional and political changes that have taken place since 2011 have, in principle, ushered in a break with the centralizing system, with the emergence of decentralized bodies, leading to a redefinition of the status and prerogatives of the governor, and sectoral deconcentrated administrations, in line with the new decentralization order.

However, it must be stressed that the public administration remains highly centralized. The deconcentrated branches of sectoral ministries, with a few exceptions, remain mere executors of decisions made at the central level, and the little autonomy they enjoy is hampered by the rigidity of budget management rules, which lengthen and slow down execution of programmed activities. In addition, these deconcentrated structures are insufficiently endowed in terms of human resources and administrative and technical capacities to properly carry out their current missions, let alone any future missions arising from the new deconcentration measures that should accompany decentralization. The restructuring of the territorial administration (i.e., the governor's role) and the deconcentrated branches of sectoral ministries has yet to begin, to align them with the new sub-national governance framework, and to prepare them to take on new missions that meet the imperatives of decentralization, namely accompanying and supporting local authorities and assuming transferable and/or shared service delivery responsibilities.

---

[21] Annual report of the High Authority for Local Finance December 2021.
[22] Idem.

6.7   Institutional and Budgetary Constraints to Overcome

Decentralization, as a policy for empowering local authorities, also faces other challenges that need to be considered in the implementation strategy. These include, first and foremost, the development of the institutional and technical capacities of municipalities (and of future regions which will have to be created from scratch). In particular, the number of senior management among total staffing is currently low, and municipalities are struggling to fill executive positions for reasons related to their financial capacity, rigid employement rules, and the lack of attractiveness of small, predominantly rural municipalities. In addition, financing of the decentralization reforms could prove costly for the national public finances which have suffered from endemic deficits since 2011. Also, implementation of the functional assignment component of the new decentralization framework will require substantial increase of the fiscal transfers, especially as the potential for mobilizing own-source revenues is relatively limited in most localities. Finally, the commitment to, not to say the ownership of, the reform process by citizens is an important factor of success. Citizens are at the same time channels for claims and pressure to stay the reform course, but also source of proposals and involvement in the administration of local affairs. The issue, therefore, is how to secure the conditions and practical modalities of effective and constructive citizen engagement; all the more so, as there is a risk that citizen involvement will lead to the exacerbation of "regionalist" claims or excessive demands that could derail the decentralization process or, worse still, exacerbate territorial disparities—the very issue to which decentralization was supposed to address.[23]

6.8   Territorial Development Policies to Put in Place

Socio-economic development efforts made since 1956 have been marred by spatial inequalities in terms of infrastructure, economic development and employment opportunities, particularly for young people and women. In fact, public policies, conceived and implemented according to a sectoral logic, have often failed to consider the specific needs of territories, have marginalized the role of local authorities, and ultimately exacerbated the territorial divide (described below). In the end, these policies made matters worse. Also, the political instability and economic stagnation observed since 2011 contributed to the aggravation of regional disparities in terms access to public services and economic development opportunities.

A World Bank report from 2014[24] provided an overview of regional disparities in Tunisia, the most important features of which are as follows:

---

[23] These tensions are already perceptible in some parts of the country.

[24] Tunisia Urbanization Review, World Bank 2014.

- *Spatial inequalities are persistent.* They have changed little over the past half century. The poverty rate in the North-West and Centre-West regions is three to four times higher than in the Centre and Greater Tunis regions. These variations reflect the large differences in average consumption, both between regions and within the same region. For all regions, the difference between urban and rural consumption was 39% on average. Similarly, the average difference in consumption between the more dynamic regions (mainly along the coast) and the lagging regions (inland) was around 29% on average, but reached 56% between the Western regions and Greater Tunis and the Central-Eastern regions.
- *Unemployment* rates vary widely from region to region, and are particularly higher in the rural interior. Unemployment is geographically concentrated in the North-West (20.3%), Central West (15.6%) and South (23.5%). Unemployment rates are lowest in the Northeastern coastal regions (12.5%). The highest unemployment rates (20–22%) are found in inland regions, compared with 7–11% in coastal regions. Moreover, in recent years unemployment has risen further in regions where it was already high, while it has fallen in coastal regions.
- *Access to basic services* (water, sanitation, health and education) is almost universal in most urban areas, but significant spatial differences persist between urban and rural areas.
- The most striking spatial imbalance, however, is in the location of *private sector activity* and employment opportunities. Private sector activity is heavily concentrated along the coast. Such concentration in coastal areas and in major urban centers is, to a certain extent, natural, given the agglomeration economies associated with it.[25] But these natural spatial patterns were exacerbated by export-promotion policies that have encouraged businesses to cluster around export support infrastructures on the coast. The spatial concentration of economic activities means that the supply of jobs is also skewed in favor of certain regions, and that densely populated pockets of the interior do not benefit from significant private sector presence and suffer from a severe shortage of employment opportunities.

The overhaul of the sub-national governance system, initiated since 2011, offers, *a priori*, the opportunity to rethink public policies with a view of promoting more inclusive, spatially diffuse, and equitable development. This requires a complete overhaul of the development policies followed to date, and the effective involvement of local authorities in the design and implementation of these policies. This also involves adopting a new regional development policy that counterbalances the promotion of territorial competitiveness (the dominant approach during the quarter century prior to the political change in 2011), with greater attention to spatial equity, and a new urban policy that enhances the position and role of intermediate cities and recognizes a more proactive role for the public sector in combating the negative effects of accelerated urbanization.

---

[25] In fact, the same trends can be observed all over the world.

The central issue for a regional development policy in Tunisia is therefore to upgrade regions that are under-equipped in terms of infrastructure and public services, and to connect people with employment locations. It also means defining and implementing a policy for cities to support their role as engines of economic growth and rural integration, while easing the negative effects of urbanization through a more balanced urban framework and strategic urban planning. But how?

It is worth noting that the issue of regional development is at the heart of the political debate in post-2011 Tunisia. The 2014 Constitution enshrines the principle of equality of living standards between regions and calls for equitable and sustainable development, and positive discrimination in favor of the most disadvantaged territories. Regional development was also considered as one of the priorities of the national development plan (2016–2020). It recognizes the importance of decentralization for the effective implementation of a "positive discrimination" strategy that fully grasps contextual differences and treats regions according to their specific needs and capacities. However, it must be said that the constitutional principles and recommendations of the national development plan have remained ineffective to date, as they have not been translated into concrete public policies and programs. Worse still, inertia and political upheavals have exacerbated regional disparities and fueled discontent, particularly in "underprivileged" regions. Similarly, the promise of decentralization to promote citizen participation and development has not been fulfilled. Yet, local authorities, if properly endowed with mandate and resources, could have made a definite contribution to the development of territories, including a greater spatial spread of employment opportunities as the Organization for Economic Co-operation and Development (OECD) points out. In a note (OECD 2018), the analysts advocate a "regional development policy, enhancing the specific assets of each region around the development of secondary urban hubs," and note that "increasing the autonomy and competencies of local authorities represents an opportunity to achieve this objective," thereby underlining the close relationship that exists between the adoption of new national territorial development policies and the decentralization reforms that can facilitate their implementation.

If this approach (OECD) is adopted, local and regional authorities will need to be put in a position to participate fully. Municipalities (as well as future regional councils) will need to be empowered to go beyond their traditional roles and assume greater responsibility in promoting local economic development and creating jobs. The State, for its part, will have a crucial role to play in initiating, stimulating, and supporting local action, while local authorities will need to be actively involved in planning local development and implementing the resulting programs and projects, either within the framework of their own competencies, or through "contracts" with the State institutions concerned, and in partnership with the private sector and community stakeholders. In the absence of such a change in the governance of local development processes, it is unlikely that the intermediate towns of the country's inland will be able to act as a catalyst for local development, particularly by integrating rural territories into the economic dynamics.

Regarding urban development, there is a concentration around three metropolitan areas (Tunis, Sfax and Sousse/Monastir), which are home to more than half the country's population. The policies pursued to date have not been able to guide or control urban sprawl, which is why all the problems generally associated with a lack of strategic management of urban growth are to be found in Tunisia, namely: costly sprawl of the urban fabric, and investment in catch-up infrastructure rather than in oriented growth; excessive consumption of agricultural land; a shortage of affordable housing for low-income people and much of the middle class; rapid development of informal housing zones (particularly in rural areas); marked underequipment of newly urbanized areas; and rising transport costs and congestion. In fact, the various urban development approaches adopted in the past have been marked by:

- Their *centralized nature* - in the sense that municipalities are not the main players in urban development, since they have few de facto functional responsibilities in this area. Indeed, the provision of services essential to urban development, such as electricity, drinking water, wastewater, and urban transport, are beyond the control of municipalities.[26]
- Their *sectoral logic*: urban management responsibilities are shared (and therefore fragmented) between multiple players (ministries, specialized urban renewal bodies or land planning agencies), with no coordinating role for municipalities.
- Their *reactive nature*—in the sense that most public interventions are not aimed at guiding urban growth, but rather at responding to emergencies, addressing malfunctions, or being under pressure from the land planning national agencies or the real estate development sector.

6.9 Conclusion

The political change that took place in 2011 was driven, to a large extent, by the socio-economic divide between the country's territories. It gave rise to new demands not only for democratization and freedoms, but above all for commitments and concrete measures to end territorial marginalization through an equitable redistribution of power and resources. Throughout the first transition period (2011–2014), political elites as well as civil society, particularly those in marginalized areas, believed that decentralization was the panacea for resolving the regional development problems and territorial divide inherited from the old regime. This explains the emphasis placed on decentralization in the 2014 Constitution, and the consensus quickly achieved in this regard, given that local authorities had hitherto played a marginal role in the design and/or implementation of territorial development policies and programs.

However, it turned out that the transition from a centralized to a decentralized and efficient system has proven to be no easy task. It will certainly take time, and will

---

[26]These services are the responsibility of State companies or sectoral ministries.

require sustained political commitment from national authorities and pressure from the territories and civil society. And it is not immune to political uncertainties, as demonstrated by the political aggiornamento of 2021, leading to the drafting of a new constitution (2022) that pays little attention to decentralization and the developmental role of local authorities.

Notwithstanding the latest (2022) constitution's narrow vision of decentralization, it is more important than ever, based on the existing legal framework (i.e., Code des Collectivités Locales), to revive the aspirations for local development and territorial equity. This means upholding and activating the role of local authorities, giving them the means to assume this role autonomously and responsibly, while clarifying and institutionalizing their "political" and financial relations with the central government, and ensuring coherence between the decentralization process and State development policies. In other words, local authorities should be put in a position to assume their development function. This requires, among other things, clear and recognized mandates, readjustment of the prerogatives and intervention modalities of the State structures[27] concerned with local development to attune them to the new legal and institutional framework of decentralization, by endowing them with authority and means commensurate with the scope of their competencies.

Ultimately, the success of decentralization will be judged by its impact on local development. Making the link between decentralization and development will not require "less State," but rather a greater commitment from a "different" State. The key questions remain unanswered, however. Does the Tunisian State, as it currently functions, really intend to foster the emergence of autonomous "developmental" local authorities, or will it continue to confine them to the provision of a few local services? And, will the State adopt policies to promote endogenous territorial development and encourage forms of central-local partnership to implement them? Or will it continue to claim exclusive responsibility for the development effort in lagging regions and implement it through compensatory programs designed and managed by the central administration? These questions remain unanswered for the time being.

References

OECD (2018) Vers une croissance plus inclusive en Tunisie. Documents de travail du Département des Affaires économiques de l'OCDE (1486)
World Bank (2014) Tunisia Urbanization Review

---

[27] In particular, the Regional Development Offices and the Commissariat Général du Développement Régional.

Chapter 7
Environmental Activism in Post-Arab Uprisings: More than Shades of Green

Zeina Moneer

7.1 Introduction

Environmental justice stresses the fairness associated with the allocation of environmental externalities. It includes equal protection from risks, active participation in decision making, and equal access to benefits as normative responses to socio environmental conflicts (Urkidi and Walter 2011; Jenkins 2018). In the Arab region, environmental movements are invoked in an environmental justice discourse, which is about privilege, asymmetric power relations, and excessive exploitation of natural resources in order to accumulate capital. The region has long suffered from a number of important and complex environmental issues. In the decades leading up to the Arab Uprisings, fears about the health consequences and other quality of life impacts, along with concerns about the justice implications of disproportionate environmental burdens, have triggered public concern and spawned environmental activism (Sowers 2012; Bullard et al. 2011).

For years urban activists mobilized around a myriad of unmet environmental needs, from a lack of clean air, inadequate waste management systems, lack of green spaces to deficient sanitation and water services. Rural mobilization revolves around a lack of access to land and water resources, hazardous waste sites, displacement and threats to livelihoods from polluting industrial facilities such as petrochemical industries (Sowers 2018). All these environmental burdens are disproportionately located near where poor vulnerable people live and thus environmental inequalities are usually built on pronounced social inequalities (Mohai and Saha 2015). For example, clean air, accessible green areas, and safe working environments are less

This study was funded by the Arab Council for Social Sciences in Beirut.

Z. Moneer (✉)
Research Institute for a Sustainable Environment, American University in Cairo, New Cairo, Egypt

accessible to low status and/or low-income groups in almost all Arab countries. The causes of these environmental injustices are multifaceted. Moreover, the overall focus on economic growth and profit is prioritized over all else across the entire Arab region (Bahout and Cammac 2018). In the following section, the main causes of inherent environmental injustice in the Arab region are presented, namely: post-colonial development and the rentier state in the Arab region; neoliberal reforms and crony capitalism, and urban inequalities.

7.2 Post-colonial Development and the Rentier State in the Arab Region

Post-colonial development in the Arab region was characterized by promotion of national manufacturing and trade protectionism, in keeping with the then prevailing nationalist sentiments (Bahout and Cammac 2018). The development model in the Arab region focused on capital-intensive reforms including land reclamation, modernization of the irrigation system through the building of dams, extraction and consumption of fossil fuels able to boost industrialization (Bahout and Cammac 2018). In the 1960s and 1970s, the rentier state model emerged in the Arab region, particularly in the oil- producing countries (Mahdavy 1970). According to Puranen and Widenfalk (2007) the rentier state has the following characteristics:

> First, only a small fraction of the population is directly involved in the creation of wealth. As a result, modern social organizations associated with productive activities have been developed only to a limited extent. Second, the work-reward nexus is no longer the central feature of economic transaction, where wealth is the end result of the individual's involvement in a long, risky, and organized production process. Wealth is rather accidental, a windfall gain, or situational, where citizenship becomes a source of economic benefit (Puranen and Widenfalk 2007, 161)

In other words, the government of a rentier state is mainly concerned with allocating the wealth among its citizens, rather than engaging in developing productive economic activities or a non-rent domestic economy (Gray 2011). Arab states usually use foreign revenue to provide well-paid public sector jobs, social welfare benefits, subsidies, and interest-free loans to win public acquiescence (Herb 2005). The distribution of shares of the oil wealth is at the core of the unwritten social contract between the state and its citizens; citizens pledge allegiance and support to the ruling elite in exchange for economic benefits (Toledo 2013).

From 1960 to 1972, due to soaring price of oil, economic growth in most Arab states was high, averaging of 10% per year. This trend was further spurred by the 1973 oil embargo (Kettell 2020). However, dependency on a single commodity and excessive reliance on hydrocarbon revenues has been detrimental, creating excess demand for non-traded goods with serious imbalances between production and consumption (Beblawi 1987). It is well-known that oil-rich countries are vulnerable to the "resources curse" whereby overall economic performance is weak and reflected in overvaluation of real exchange rates, large fluctuations in inflation rates, and underdevelopment of human capital (Gylfason 2001). Under this premise, while oil-rich countries in the Arab region received large rents in the 1970s and 80s, their economies performed poorly as compared with countries that were net oil importers during the same period (Rutledge 2013). Furthermore, rentierism does not only effect the State but the entire society, where patronage is relied upon and unearned income preferred (Hameed 2020). The flow of external rents and their distribution based on nepotism perpetuates the monopoly of the ruling elites over the economy and discourages citizens from playing an active role in the economy, which aggravates socio-economic inequalities and stifles economic growth (Tullock 1989; Mihalyi and Szelenyi 2019).

In the 1980s a dramatic fall in oil production and price plagued the oil-rentier economies of the region. Its oil production decreased from of 31 million barrels per day in 1979 to 18 million in 1982, while the price of oil fell by 50% (Özekin and Arıöz 2014). With the slump in oil prices, oil-rentier states faced a dilemma. On the one hand, they could not provide their earlier rates of rent distribution, and on the other, in order to protect their legitimacy and avoid social unrest, they tended to borrow heavily from the international community (Özekin and Arıöz 2014; Benli Altunışık 2014). The rentier economies of the Arab region were left with serious fiscal deficits and accumulated debts (Cooley 2001). For example, Between 1984 and 1991 the Current Account Deficit (CAD) of Saudi Arabia rose from US$18 billion to US$27 billion. In Egypt, low oil prices coupled with inefficient allocation of resources and state-led industrialization led to budgetary deficits and increasing external debts which reached US$39.8 billion in 1986 (Hameed 2020). Reliance on loans to meet pre-existing social obligations proved to be only a temporary solution, particularly given the growing population in the region. Eventually, its rentier states had to adopt austerity measures and to implement major cutbacks in government spending and subsidies which were necessary to cushion the effects of oil price slumps (Roberts 1986). Therefore, significant segments of the populations in the region found themselves with poor basic sustenance, a condition worsened by inequality that cuts across every sector of life, including the rural-urban divide, class, income and environmental stresses (Khouri 2019).

7.3 Neoliberal Reforms and Crony Capitalism

Facing high population growth, declining per capita income, increasing levels of unemployment, and accumulation of external debt (El Erian and Tareq 1993), Arab countries have gradually sought to address their economic problems by implementing economic liberalization and structural adjustment (Murphy 1998). These adjustment and economic reform programs were designed and implemented based on International Monetary Fund and World Bank policies, policies that enshrined "free market" principles restriction of the ability of State to regulate the economy (Aarts 1999). According to Alissa (2007, 3), "reforms focused on four main areas: cutbacks in government spending, privatization of state-owned enterprises, reduction of barriers to trade, and liberalization of interest and exchange rates." However, the pursuit of economic liberalization without also implementing significant political liberalization and meaningful democratic transformation failed to meet employment needs and secure social services for the growing population of the region (Afouxenidis and Kourtelis 2017). Furthermore, the private sector has not become truly competitive while privileged groups who enjoyed close ties with the ruling elites controlled basic utilities and captured much of the gains of the privatization and investment opportunities (Owen 2015; Malik 2019). In addition, social welfare programs have been stretched, aggravating the sense of insecurity and alienation among vulnerable population groups faced with unemployment, decline in government spending on social services, and marked deterioration in living standards (ILO 2012; Sika 2012; Hanieh 2015).

As a result, social inequalities have deepened, and rates of poverty have risen. States are unable to sustain the types of subsidies that used to placate poverty-stricken rural and urban populations (Owen 2015). Moreover, the implementation of liberalization policies in the early 1990s ended growth in public-sector employment. This hurt the middle class hardest. It had formed the main economic group in the Arab region, supporting its regimes. This has caused the middle class to gradually shrink to about one-third of the Arab population by 2010 (Hertog 2023). In addition, although according to the standardized measure of poverty (the $1.9 per day international poverty line), poverty has declined in the Arab region since 1990, the Arab region fell short of reducing extreme poverty by half from 1990 and vulnerability to poverty remains high (Abu-Ismail and Al-Kiswani 2018). These findings conform to the outcomes of multidimensional poverty surveys by the UNDP and OPHI, which evaluate deprivation in health, education, access to water and overall living conditions. In 2018, the Arab Multidimensional Poverty Report estimated that nearly one fifth of the Arab population is extremely poor and two thirds are either poor or vulnerable to multidimensional poverty (OPHI 2018; Abu-Ismail and Al-Kiswani 2018).

Although the Arab countries are rich with natural resources which underpin their economies, their revenues in the context of captialism have been linked to poor development outcomes, inequality, and poverty (Patrick 2012). Furthermore, these resource revenues are usually dominated by a small corrupt elite or a few institutions where there is the line between public and private capital is porous (Sika 2012). This presents a fertile field for patronage networks and clientelism to thrive in a way that institutionalizes rent-seeking behavior, curtails the establishment of mechanisms for reallocation of income flows, and sharpens income inequality (Assouad 2020).

Furthermore, these resource revenues tend to benefit the urban middle classes, and the coastal and urban centers, while the periphery or rural areas lag behind and bypassed by development policies, creating significant spatial inequality (Sika 2012). It is estimated that during the 2010–20 period, the richest 1% of Arab citizens increased their control over regional assets from 31.4% to 44.9%, while the richest 10% of Arab citizens increased their control over regional assets from 69.7% to 81.1% (Abu-Ismail and Hlasny 2022).

What aggravates the inequality is the fact that rural and peripheral areas are home to the vast natural resources that generate vast wealth, while the local communities in these regions see little of the economic benefit. Conversely, urban centers and cities are the sites of state-led development. Their populations are usually better off and have better access to public services (Mills and Alhashemi 2018). Tunisia offers a clear case of spatial disparities as the country has become mired in a skewed "metropolis-satellite" relationship. Its center exploits capital and resources from its peripheries, reinforcing economic marginalization of the peripheral areas and worsening social and income inequalities across the country (Sadiki 2019). For example, although the southern governorate of Gafsa has the largest reserves of phosphate minerals, the local people reap few benefits from phosphate export revenues, bear the costs of environmental pollution, and suffer from high rate of unemployment (Al Jeezrah 2020). Another illustration of the urban-rural inequality that marks different regions of Tunisia relates to access to fresh water. Despite being home to the largest share of freshwater supply in Tunis, the North-West suffers water shortages in the summer months, when the government cuts water supply to rural and interior regions and redirects it to the capital and coastal cities as a way to support socioeconomic development (Pope 2016).

Similarly, when Sadat initiated liberal market reforms in Egypt in the 1970s, primarily through the policy of Infitah, the policy shifted from aiming to attract investment and reinvigorate the productive capacity of the economy, to a general mode of rampant consumption characterized by consumerism (Shechter 2008). Furthermore, rapid urbanization and encroachment onto agricultural land around metropolitan Cairo and Alexandria, as well as around other cities in the Nile Delta, have reportedly increased pressure on urban infrastructure, basic service provision, and agricultural productivity (Finaz 2015, 2).

Neoliberal economic reforms were further pursued during the Mubarak era, and particularly the privatization experience since 1990s, resulting in the rise of "networks of privilege" and "crony capitalists" that enjoyed close ties with the State in ways that enabled them to obtain special crony privileges and to bend or break planning laws and other legal constraints when it suited them, including environmental guidelines (Henry and Springborg, 2010). Examples of these privileges include: accessing scarce natural resources (e.g. arable land) at cheap cost, capturing fuel subsidies which are crucial for the profitability of energy intesnive sectors such as iron, steel and cement, winning lucrative business deals such as large-scale public construction contracts (Chekir and Diwan, 2013a, b). Furthermore, well-connected business cronies benefited from selective implementation of government rules and regulations such as generous tax advantages and less restrict loan and credit conditionality (Checkir and Diwan, 2013b). For example, in the last year of Mubarak's regime, 92 per cent of private sector loans were obtained by business elites at the expense of smaller and less connected businesses (Diwan et al., 2013). These accumulated privileges rewarded to crony businesses have not necessarily resulted in an increased investment in innovation or efficient environmental performance (Diwan et al., 2013). For instance, the business elites during Mubarak's era operated primarily in energy and infrastructure benefiting from favorable access to capital and tariff exemptions and achieving 60% of net profits, while employed only 11% of the labor force and leading to increased pollution and environmental harms (Kirsanli, 2023).

7.4 Urban Inequalities: Gated Communities Versus Slums

The Arab region is one of the fastest urbanizing regions in the world (Sharp 2018). In the 1960s, the urban population in the Arab region was estimated to be 35% of the total regional population according to the World Bank. Currently, 64% of the population of the Arab world lives in cities and urban centers, exceeding the global average of 55% (World Bank 2021a). It is expected that the population of cities and urban areas will increase to 75% by 2050 (Elgendy and Abaza 2020). In addition to natural demographic increases, the growing concentration of the population in major cities and towns over the past few decades is primarily attributable to pull factors drawing residents from rural to urban areas (i.e., better employment opportunities and service provision) (Smiley and Emerson 2020). Urbanization is far from uniform across the region. In Egypt for example, urbanization has been slow over the last 70 years, rising modestly from 31% in 1950 to 43% in 2020. In Lebanon by contrast, urbanization has increased much more rapidly from 42.3% to 88.9% during the same period. And unlike Egypt, Lebanon's urbanization did not follow a linear trend, but spiked in wake of regional and national wars in the 1980s, 1990s and 2000s (World Bank 2021b). The rapid urbanization process is reflected in the proliferation of urban megaprojects, gated communities, retail malls, airports, ports and highways in the Arab region. In this regard, from 2006 to 2016, cement production almost doubled in the region's major cement producing countries, such as Saudi Arabia (from 27 to 61 million tons), Egypt (29 to 55 million) and Turkey (47 to 77 million) (Sharp 2018).

In the Arab region however, the concentration of the population in large urban agglomerations has not addressed the scarcity of resources through improved efficiency. Instead, rapid urbanization has exacerbated the disparity between population growth and the scarcity of resources and services. The transformation to urban life and concentration of investments in specific economic sectors (i.e. tourism and real estate) have further escalated urban problems of uneven territorial development, land speculation, urban fragmentation and social inequalities (Elgendy and Abaza 2020). In this regard, only rich can afford to segregate themselves in small, gated communities while the poor are crammed into under-equipped shanty districts (Adham 2005). According to Abaza (2006), gated communities are the most visible outcome of neoliberal adjustments in a number of Arab countries (Egypt, Tunisia, and Morocco). This has enriched national and foreign investors while creating a physical segregation in the fabric of Arab cities. In Cairo, for example, upscale gated communities are burgeoning on the outskirts of the metropolis, catering to the upper middle class which seeks an elite, "safe," controlled lifestyle and to escape the polluted metropolitan environment (Hendawy and Saeed 2019). On the other side of the urban spectrum of Arab cities, slums or informal settlements are characterized by insecure land tenure, poor housing, pollution, health hazards, high crime rates, and a lack of such basic services as water, electricity and sewage (UN-Habitat 2003). Furthermore, slum areas are often the recipients of the city's nuisances, including industrial effluent and noxious waste, and the only land people are able to build houses on is often physically deteriorated, dangerous, or polluted, effectively land no one else wants to inhabit (Farid 2019). For example, the Cairo slums of Manshiet Nasser and Duweiqa were categorized as unsafe zones due to their exposure to industrial pollution and being located on areas that are subject to flooding or rockslides (Farouk 2020). In addition, with the on-going upheaval in some parts of the Arab world such as Syria, Libya and Yemen, informal settlements have become the main recipients of refugees and Internally Displaces people. More than one third of Syrian refugees live in informal settlements in the Beqaa Valley in eastern Lebanon. Inhabitants of these informal settlements suffer from poor housing quality, precarious environmental conditions, and health risks (Naggar 2020).

While it is true that there have been great efforts in order at environmental reform in the Arab region, this legislation, implementation and enforcement has mainly focused on the risk management and containing public opposition to environmental hazards, rather than on the elimination of harmful production or on unsustainable extraction and use of natural resources. Furthermore, the budgets allocated to environmental programs are well below 1% of GDP for any of the countries in the region. A further sign of weakness is that funding for these programs is reliant on external agencies (Tolba and Saab 2008). Large-scale economic development projects are not currently preceded by sufficient and transparent studies of their environmental impacts. In addition, authorities in charge of the execution of environmental laws often do not coordinate with other relevant authorities such as defense departments, and agriculture and water resource agencies, contributing to non-compliance. The lax environmental laws and lack of compliance with environmental regulations results in environmental degradation and pollution while those most negatively affected by industrialization, rapid urbanization and pollution remained marginalized.

7.5 Environmental Activism in the Arab Region: Categories, Frames and Tactics

The intensifying environmental and health threats of the past few decades have given rise to environmental activism. Environmental campaigns have emerged as the most publicized and flexible forms of activism in the Arab region. Environmental activists have learned from other international mobilization campaigns how to use collective action frames, establish solidarity networks, build collations across scales, and combine tactics of moral suasion, lawsuits in the courts, and direct-protest actions designed to obstruct and to draw attention to environmentally harmful policies and projects. Other strategies included public-education and media campaigns, and conventional lobbying of policy makers and political representatives (Williams 2013; Sowers 2018). However, environmental campaigns in the Arab region often gradually lose momentum as they lack the financial resources they need to deploy for a given campaign, their claims are undermined by state suspicion and repression, and their efforts are hindered by the power of bureaucracy and capital (Sowers 2018).

Environmental activism in the Arab region falls into several categories. The first category tends to work mostly on short-lived campaigns directed against a local source of pollution (i.e., a factory or an incineration plant) or against plans to erect new infrastructure near inhabited settings, particularly in urban areas. Participants in this category tend to be residents of the district or neighborhood who organize through social networks and then dissolve once their cause is addressed (Davydova 2021). An example here is the environmental campaign to contain the mercury pollution in the Oued El Harrach river basin in Algeria, which was caused by industrial wastewater (Yoshida 2016).

The second type of campaign tends to focus on issues that are absent from the governmental agenda: such as recycling and urban greening (Davydova 2021). An example here would be the Tunisie Recyclage movement, which created a system to recycle garbage in Tunisia's capital (Lageman 2016). These types of groups rarely engage in protest activities and tend to focus their energies and resources on collective neighborhood organization, active participation in neighbourhood redevelopment projects, lobbying and engaging the general public though awareness campaigns (Anguelovski 2015).

The third genre of environmental activism in the Arab region focuses on public monitoring and oversight of environmental governance and urban policy at the national and local level. Watchdogs might also provide alternate estimations of environmental data (especially when data is not declared or reliable) or coordinate campaigns for access to environmental data, demanding transparency, and accountability particularly of corporations and industries (Davydova 2021). An example here is the grassroots initiative to create alternative, civic-based monitoring of the solid waste management in Lebanon. The initiative is coordinated by the Waste Management Coalition and Human Rights Watch (Human Rights Watch 2020).

In order to raise awareness and attract wide public support for their causes, environmental campaigns in the Arab region used to publicize adverse local environmental and health impacts that affect communities in a direct physical way (Sowers 2012). This way of framing environmental campaigns in the Arab region made the environmental issues easier to grasp by the public, more intuitively morally wrong and more likely to trigger feelings of resentment among targeted groups, and eventually more likely to mobilize participation and trigger solidarity (Green 2018; Moneer 2021). The injustice frame imbues discourses about environmental campaigns in the Arab region. Injustice frames invoke that, in addition to essential human rights of clean air, land, and water, etc., people have a right to be free from human experimentation and environmental hazards (Taylor 2000).

In addition, activists tended to articulate their interests and goals in contradistinction to elites and other opponents. For example, foreign multinationals are usually portrayed as the main culprits of pollution and described as profit seekers at the expense of peoples' health and rights to a safe and clean environment (Moneer 2022). Environmental injustices that can be assigned to deliberate actions of identifiable actors are particularly effective in building advocacy and solidarity and in stripping opponents off their social license (Keck and Sikkink 2014). The transformed communications landscape of the past few decades has facilitated coalition-building across social and geographical divides between environmental groups and other actors who might have different objectives (Berriane and Duboc 2019). For example, collective protests in 2008 in the mining region of south Gafsa, Tunisia, focused on environmental injustices and socioeconomic concerns such as rising food prices, intensified exploitation of natural resources such as oil, phosphate and ore and unequal distribution of benefits, lack of basic services, and increased marginalization of the region (Ayeb 2011). This coalition brought together marginalized populations, such as small peasants, miners and urban poor with established elites, such as urban based lawyers, and environmental and human rights activists (Allal and Bennafla 2011).

Similarly, the privatization of Jordan's mineral industry in the mid 2000s—part of the government's accelerated neoliberal reforms—resulted in widespread public resentment against what was described as an illegitimate appropriation of Jordan's natural resources, and by extension, an abrogation of the state's redistributive obligation (Al Rawashdeh and Maxwell 2013). Consequently, a diverse cross-section of social movement and constituencies, spanning environmental, labor, and rural community activists, informed transgressive mobilization practices and discourses targeting corporatization and privatization of the natural resources and local people's dispossession of their national assets (Lacouture 2021). This is particularly due to the popular perceptions that natural resources belong to Jordanians as part of a historical social contract where the state is responsible for the (re)distribution of social and economic benefits. The abrogation of this "contract" through illegitimate privatization represented a moral violation (Lacouture 2021; Loewe and Albrecht 2022). In an active effort to unify labor and popular grievances and concerns, the notion of privatization was introduced as the main driver of corruption, unemployment, the ills of the neoliberal extractivist system and the unfair distribution of the revenues related to the exploitation of Jordan's national natural resources (Lacouture 2021).

Activists and environmental defenders in the Arab region used to denounce and resist mechanisms of domination and dispossession in a direct way, rather than relying on lobbying or media outreach. This was particularly the case of energy and resource-intensive activities such as big dams, mining, energy intensive agriculture and industries (Dwivedi 1998; White 2013). They usually adopted pragmatic attitudes and their actions often aimed at addressing specific environmental controversies, often doing so one at a time. These characteristics are closely allied to tendency to deal with effects rather than structural causes of environmental ills and without significant lifestyle changes or threats to economic growth (Davydova 2021). This is particularly true in North African countries where the architecture of capitalist markets does not provide any mechanisms to counter the environmental hazards that large-scale production and consumption processes inflict on the environment. In these countries, the State is subservient to the bindings of the influential business groups that need incentives to ensure that they remain profitable—regardless of environmental considerations—to ensure growth and ample employment opportunities that are needed to provide for the regimes' legitimacy (Movahed 2016).

7.6   State Responses to Environmental Activism in the Arab Region

The regression of democratic values in many countries, including across the Arab region, and the restriction of civic space, have created a negative environment for environmental activists (Sowers 2018; El-Mikawy 2020). Criminalization and stigmatization are common strategies used to weaken activists and stifle their ambitions (Rodríguez-Labajos et al. 2019). This criminalization is associated with the introduction of legal changes in many countries to create tighter controls of public space. In this regard, legal systems are used systematically to inhibit protests by, for example, denying permits for activists to organize mass protests. In addition, non-violent actions such as marches, roadblocks, and sit-ins, which have historically been adopted by a variety of movements, have been identified as crimes (Rubiano 2021).

Stigmatization and smear campaigns against environmental activists are enacted through the dissemination of false information that creates a negative perception about the environmental activists among the public. Stigmatization campaigns are carried out by different groups, including corporations, non-state anti-rights groups and senior state officials who cast doubts about the intentions of the environmental activists and portray them as anti-development, destabilizers of democracy or agents of foreign influence (Sowers 2018).

Moreover, smear campaigns are often supplemented by legal and extra-legal restrictions on activists and the functioning of their environmental organizations (Rowlands and Peña 2019). Another striking element of the responses to environmental mobilization is the well-documented and now much-discussed phenomenon of closing space for civil society in dozens of countries—nondemocratic but also democratic—around the world (Carothers and Youngs 2015). These efforts to choke off civil society usually include an array of formal and informal measures, ranging from restrictive NGO legislations to tactics to undermine the public legitimacy of civil society actors and consistant measures to restrict their funding resources and impede their operational effectiveness (Brechenmacher 2017).

Given the restricted political environment, limited available resources and looming threats to environmental activists, environmentalists in the Arab region often accept compromises and concessions from both the authorities and industry (El-Mikawy 2020). Electoral politics and mobilizing an organized Green political movement have played second fiddle to protests, litigating, and negotiating with government agencies and corporations in the Arab region. In Europe, Green parties play a key role in bringing the influence of the grassroots environmental movement directly to government, make the environment a central concern of public policy, rendering the institutions of the state more democratic and transparent, and making corporations more accountable (Elliott 2017). In the Arab region, grassroots environmental movements failed to develop into broad-based political parties capable of winning elections and serving at the highest levels of government. One reason could be the fact that most civic movements unfold under tight state supervision and sometimes repression (Shalby 2016). This has led to the fragmentation of associational life and to difficulty in creating formal organizations such as political parties. Civic movements have become mainly concerned with their own survival (Brechenmacher 2017) or achieving their narrow results-based objectives. In addition, environmental movements have largely failed in forging coalitions and partnerships with other civic movements and/or established political parties. Interconnected and jointly coordinated civic movements are more capable of confronting State repression and play an active role in contentious politics. However, individualized movements are more likely to be controlled and suppressed (Brechenmacher 2017). In addition to the political restrictions, legal and administrative regulations put similar restrictions on the formation of political parties, their registration, and operation (Ottaway 2021a, b).

For example, Egypt's Political Parties Law #40/1977—which was the major constitutional and legal framework that governed the formation and the organization of political parties in Egypt in the 1977–2011 period (Human Rights Watch, 2007), established the Political Parties Committee (PPC) which had great leverage to refuse the registration of new parties, to freeze existing parties' licenses, to suspend parties' activities based on subjective and vague criteria, and to ask Cairo's Supreme Administrative Court to dissolve parties and redistribute their funds (State Information Service 2023). In this regard, the PPC refused the registration of the Egyptian Green Party, claiming that the party did not have a distinctive programmatic identity. However, this claim was refuted and the party was legally established by a ruling court in 1990 (Fahmy 2002). Even those Arab Green Parties that have managed to get registered and obtain legal status have not managed to achieve significant electoral gains given the fact that they lack the resources to build efficient organizational capacities and to develop grassroots constituencies. The parliamentary election results in Egypt exposed the weakness of the Green Party and its inability to provide candidates for all the parliamentary elections before the 2011 revolution (Hyde 2011). Following the revolution, the Green Party was only represented in the consultative council (Shura council) in 2013. In Tunis, there are two green parties. The Green Tunisian Party was established in 2004 but remained illegal under the rule of Zine El Abidine Ben Ali. It was only legalized in the wake of the Tunisian revolution.[1] (Wikipedia 2021a). A second party, the Green Party for Progress, was established in 2005 and legalized in 2006. In the 2009 general election, it won 6 seats in parliament with 74,185 votes and a 1.67% vote share.[2] It got only one seat in the

---

[1] Anonymous interview 1.

[2] Anonymous interview 2.

2018 municipal elections, with 150 votes (Dejoui 2018). In the 2022 Tunisian parliamentary elections, the Green parties failed to elected a single member.[3] In Morocco, there are three Green parties which are: the Environment and Sustainable Development Party established in 2010 (York 2010), the Green Left founded in 2005 as a splinter from the Unified Socialist Party, and the Moroccan Ecologist Party-Izigzawen which is the new self-denomination of the Moroccan Amazigh Democrat Party.[4] The three parties have no representation in the parliament.[5]

7.7   The Arab Uprisings and New Political Opportunities for Environmental Activism

In the wake of Arab uprisings, numerous environmental movements emerged across the Arab region which reflected subnational disparities and inequalities as well as changing opportunities for activism (Moneer 2021). These facilitating opportunities—driven by the spread of alternative media and the rise of new generation of activists eager to experiment with new forms of networking—coupled with intensifying grievances in the regions fueled by the deteriorating socioeconomic conditions, repression and corruption, have led to the widespread political mobilization around environmental issues, problems that had remained uncontested and unaddressed before the Arab Uprisings.

The conditions which have facilitated the trend include: the rising role of non-elite social agency, the widespread use of social media and social networking, the resuscitation of civil society, and the emergence of its transnational extension. The Arab Uprisings have placed more emphasis on the political role of non-elite social agency and created crosscutting ties within previously highly fragmented and localized political opposition movements (Bayat 2013). During the heat of the 2011 protests, networks were formed among the previously disparate groups of workers, labor organizations, urban poor and small farmers, students and local self-help initiatives (Bayat 2013; Achcar 2013), fostering the development of a "new dynamic and inclusive political culture" among oppositional activists (Abdelrahman 2011).

Another facilitating condition that emerged in the immediate pre-Arab Uprisings was the increasing use of digital media and social networking in bringing decades of shared grievances to fruition in virtual and public displays of collective behavior (Carty 2015). The expanding use of social media drew public discussion away from state monopoly on mainstream and traditional media, provided an information tool and allowed for coordination of social movements and increased the participation of underrepresented populations (Salanova 2012; Carty 2015). This is particularly true given the nature of digital media and social networking and their effects on collective

---

[3] Anonymous interview 1.

[4] Anonymous interview 3.

[5] Anonymous interview 4.

mobilization where activists, for example, can develop networks that are loose, leaderless, horizontal and allow for different political orientations described as ideologically flexible, hybrid, and cosmopolitan. These features became decisive in shaping the Arab uprisings (Kingston 2019; Karduni and Sauda 2020). In turn, these new styles of networking and mobilization forged shifts in political opportunity structures in the region. According to Benkler (2006), social media outlets have not only reduced the cost of producing and publishing media content, they have decentralized media production, making it much harder for repressive regimes to prosecute activists and censor media outlets.

A third facilitating condition is the resurrection of civil society, particularly with an initial opening of the political systems resulting from regime changes and a wider trend of democratization processes in the region (Plaetzer 2014). The Arab uprising managed to disrupt existing structures of powers and to reconfigure state-society relations as new actors came to the fore, including political parties, newly formed political coalitions, and previously marginalized civil society organizations (Sadiki et al. 2013; Harders 2015; Heydemann 2016). In addition, the driving force behind the Arab Uprisings did not come from what can conventionally be defined as formal organized civil society. The trigger instead came from the realm of informality—from everyday citizens who, connected by technology and united by shared grievances, managed to challenge the status quo in their respective countries to a degree that civil society organizations had rarely alone accomplished (Yom 2015). The result of the demonstrable political power of the people was an intellectual rediscovery of the concept of civil society, a transcendence of the constraints of national borders and an awareness of its transitional character (Kostovicova and Bojičić-Dželilović 2013). Protesters reached across borders, both within the Arab world and outwards to other regions. New political mobilization tactics were learnt. New ways to circulate information, to generate common norms of solidarity, and to publicize to outside audiences the repressive actions of governments were adopted. These external linkages reconceptualized the notion of civil society. Today, while civil society will always respond first and foremost to local concerns, activists will adopt innovative repertoire of contention and resort to mobilization practices that are applied beyond their national borders and domestic confines (Yom 2015).

The following section focuses on three environmental movements that flared up in a number of Arab counties in the wake of the Arab Uprising, pinpointing the main claims of these environmental movements, their broader political, economic and social underpinnings, and their final outcomes.

7.7.1 Algeria's "No to Fracking" Movement

Since 2013, the politically marginal but rich oil-producing region of the Algerian south has witnessed successive waves of unrest against the plans of the authorities and multinationals corporations to frack for shale gas (International Crisis Group 2016). The remote Saharan town of Ain Salah (pop. 35,000) in the Ahnet Basin became the site of recurrent demonstrations since the state energy giant Sonatrach announced it had successfully completed its first pilot drilling in the region in December 2014 (Longeray 2015). In January 2015, protests erupted in the main square of Ain Salah before spreading to neighboring towns, including Tamanrasset, Adrar, and Ouargla (Boersma et al. 2015). Concerns over health—such as risks from proximity to fracking wells, and contamination of water resources—were the most significant issues voiced from the outset of the unrest. Other concerns raised regarded the potential threats the fracking process posed to local livelihoods (Longeray 2015). The Sahara is usually portrayed as a spacious land that is sparsely populated, however is rich in fossil fuel resources. Therefore, the Sahara is a perfect location for multinational fossil fuel companies to get a hold over fossil fuel resources to provide Europe with its increasing energy demands that are needed to sustain their extravagant consumerist lifestyle (Hamouchene 2022). However, this narrative overlooks the questions of ownership and sovereignty that were prominent in the anti-fracking campaign with the activists' recurrent confirmation on their right to defend their "home" from a mode of development imposed by outside actors (Hamouchene and Pérez 2016). The anti- fracking highlighted the opposition of the people of Sahara to the practices of extractivism that facilitate the plunder of natural resources by profit-seeking multinational companies, dispose the local communities and turn their land into sacrifice zones where they are stripped off their livelihoods resources and are exposed to dangerous health and irreversible environmental harms (Hamouchene 2022). The fracking concession evoked the sentiments that are related to the 1950s nuclear testing in the Algerian Sahara (Hamouchene and Pérez 2016; Belakhdar 2020). This period of nuclear testing in Southwestern Algeria was noted for the absence of any safety precautions for the indigenous Touareg population, many of whom were exposed to nuclear radiation. Moreover, although France recognized Algerian independence in 1962, France was able to negotiate a secret agreement that allowed it to continue with underground nuclear testing until 1966 (Crapanzano 2010). When France abandoned its Sahara test sites' facilities, it did little to clean up the residual contamination, and even less to inform the local population about the existence of dangerously radioactive zones close to their villages (Jacobs 2013).

The anti-fracking movement was led by the National Committee for the Defense of the Rights of the Unemployed in the context of broad discontent in the south of Algeria, which suffered from long-term political marginalization, economic exclusion and underdevelopment despites its oil riches (Belakhdar 2020). The popular mobilization against fracking in Ain Salah was linked to demonstrations in other southern regions suffering from chronic regional inequality and economic disparity (Hamouchene and Pérez 2016). In effect, inequality between Algeria's northern and southern regions are significant. The South represents 87% of Algeria's total land area and is where the fossil fuels that are the main source of the government's income are located. Yet it suffers from high rates of poverty (double the national rate), sorely lacks basic infrastructure, and unemployment is endemic among youth (unemployment rate is 50% while the national rate is 20%) (Boersma et al. 2015; Aziz 2019; Borgen Project 2023)

By raising the questions of who benefits from the state's natural resources and who decides how these resources should be used, the anti-fracking movement managed to highlight the intersection of environmental questions with issues of regional (in)equality and state-society relations, bringing it to the forefront of the public sphere. In addition, the activists managed to bring an anti-imperialist dimension to the anti-fracking movement. They emphasized the fact that while France has banned Total and other companies from gas fracking on its territory, it was promoting it in its former colony Algeria (Hamouchene 2020). The EU continues to push for more liberalization in energy contracts despite the fact that Algeria's current Hydrocarbon Law maintains Sonatrach' monopoly over energy concessions and ensures its majority-stake engagement in any energy venture. Foreign investors see this as limiting their profits (Boersma et al. 2015; Escribano 2018). By emphasizing the double standard, the activists anchored their resistance to the issuing of fracking concessions in the Sahara in a broader opposition to the modern colonialist project of neoliberalism, which reproduces colonial relations through exploitation of cheap natural resources, imposes neocolonial structures of global trade, and exacerbates pre-existing social and economic injustices afflicting vulnerable and marginalized local communities (Moneer 2022).

After a year of consistent mobilization in a number Algerian cities, and despite numerous attempts to contain the movement, the government had to yield to the demands of the peoples of Sahara. It declared a halt to drilling activities in May 2015 (Belakhdar 2020). Public dissatisfaction with the government's approach to quell the protests contributed to former Energy Minister Yousfi losing his position in May 2015 (Brahimi 2015). This outcome is considered a success in a country that relies heavily on hydrocarbon revenues to provide basic services to it citizens (Belakhdar 2020). More importantly, the anti-fracking movement managed to create solidarity among the oppressed and marginalized, be it with the indigenous communities of Sahara region, or with the unemployment movement, triggering nationwide opposition to the corruption and repression of the political ruling elites (Kinniburgh 2015; Werenfels 2016).

7.7.2   Lebanon's "You Stink" Movement

During the summer of 2015, thousands of Lebanese protested Beirut's months-long garbage crisis (Chayes 2018). The crisis started when a group of local activists demanded the permanent closure of Beirut's main landfill, Nameeh, and refused the Environment Minister Mohamad Mashnouk's plan to extend operations in the landfill until late January 2016 (Atwood 2019). The protests led to the closure of Naameh landfill without the provision of an alternative site as final destination for Beirut's solid waste (Abu-Rish 2017). In parallel, Sukleen—the main company responsible for waste management in Beirut and Mount Lebanon—stopped trash collection due to the expiration of its contract (Kerbage 2017).

At the time of the crisis, Greater Beirut produced 3000 tons of garbage per day. Within a week of the beginning of the crisis, the Minister of the Environment estimated that 22,000 tons of garbage was uncollected from Beirut's street (Issa and Zaatari 2015; Wood 2015). Frustration with the crisis spawned a movement—called "You Stink"—as a response to the government's inability to find sustainable solutions (Kerbage 2017). During the eight months of the garbage crisis, 'You Stink' continued its mobilization, successfully calling for cross-sectarian protests that attracted people from all walks of life (El Ksayer 2017). At the start of the crisis, specific solutions were framed in purely scientific and technical terms. The discourse was framed as an issue of technical and scientific expertise as a way to legitimatize the claims of the actors concerned and to emphasize opportunities that sounder and more sustainable environmental practices might offer (Eder et al. 1995). As the garbage crisis persisted, and as the government responded to the protests with violent, the way the crisis was framed was transformed. Activists were able to link the trash crisis to previously dissociated themes such as sectarian politics and corruption (Tadamun 2016). In this way, the garbage crisis was portrayed as a part of the wider national problems of corruption and blatant efficiency of the sectarian political system (Civil Society Knowledge Center 2014).

The accusations of corruption in the waste management sector, as in other public services, are substantial. Such public services are operated by private businesses owned by powerful political leaders who facilitate tender violations while promoting deregulation and privatization of public service to maximize their profits (Abu-Rish 2015). In the 1995–2015 period, the waste management sector was dominated by two sister companies, Sukleen and Sukomi, responsible for street cleaning, trash collection, recycling and dumping waste in defined landfills (Tadamun 2016). Since 1995 Sukleen had become the main waste management provider in Beirut and Mount Lebanon. Sukleen was founded by Maysara Sukkar, who has strong business ties with the political leader Rafiq Hariri (Abu-Rish 2015). Sukleen managed to monopolize trash collection services through a contract the Council for Development and Reconstruction (CDR). The contract was renewed three times without the open tender process to would have allowed competing companies to bid (Chaaban 2016). From handling 800 tons of garbage per day in 1994, by 2015 Sukleen was handling 2600 tons from about 400 municipalities (Chaaban 2016). With each contract renewal, trash collection and dumping fees increased. The fee was set at $45 USD per ton in 2015, even though the international average for waste management fees at that time was only $11 per ton (Abu-Rish 2015). The waste management fees were paid by the Independent Municipality Funds (IMF), created in 1979 to allocate funding to local municipalities. However, these fees which filled the coffers of private businesses deprived municipalities of the IMF funds they needed to operate their other services, leaving them underfunded and incapable of fulfilling basic services (Atallah 2015).

In light of the widespread corruption of the political elites and the lack of transparency and inefficiency of the waste management sector, the 'You Stink' movement engaged in fighting the larger battle, not confining its demands simply to the symptom of corruption stinking on their doorstep. The campaign transformed its demands from merely demanding a solution to the garbage crisis to a radical reformation of the entrenched, sectarian, political system (Patel 2018). Although the movement's adherents did not provide a vision of an alternative political system, they did assert the need to reform the electoral law to allow for proportional representation. Following the enactment of a new electoral law, the thrice-postponed parliament election (2013, 2014, 2017) was held on May 6, 2018 and a new president elected (Yacoubian 2018). These were the first elections in nearly a decade.

In parallel, the government recognized how untenable the garbage crisis had become and enacted a number of changes that responded to the movement's demands, including appointing Akram Chehayeb, Minister of Agriculture, as chair of a garbage crisis committee established on the 31st of August 2015 (Strobl 2016). Chehayeb developed a four-year plan that entailed re-opening the call for bidding for waste management in the "Sukleen region" and building two coastal landfills in the Beirut suburbs of Costa Brava and Bourj Hammoud. The bids were to be awarded by the Council for Development and Reconstruction (CDR) (Civil Society Knowledge Centre 2016). In March 2016, the Council of Ministers approved the Chehayeb plan and the city's waste management companies began relieving Beirut of its garbage problem, removing piles of trash from the streets and bringing them to the two new dumps (Strobl 2016).

Although the 'You Stink' movement did not manage to force long term sustainable solution for the trash crisis, it did achieved some victories. The Naameh landfill was closed, the Minister of Environment resigned from the government committee on waste management, and the government acknowledged that municipal authorities should gather the trash (Atwood 2019). Still, the most lasting achievement of this environmental movement has been the networks and alliances that emerged during the course of popular mobilization, along with the public airing of a whole set of questions previously hidden in the folds of sectarian politics and entrenched patronage linkages (Abu-Rish 2017). More importantly, the 'You Stink' campaign resulted in the rise of new parties such as "Beirut Madiniti" (Beirut is my city) that have challenged the well-established political parties in municipal elections, winning 32,000 votes (40% of voter turnout) in the May 2016 municipal election. Due to how consociational political system is designed, these votes could not secure any seats in Beirut's municipal council. Yet, this level of support was very encouraging for a nascent political party (El Ksayer 2017). Even more importantly, the mobilization practices and the language used by Beirut's You Stink movement of 2015 provided entryways for the nationwide protests against corruption that erupted in late 2019 (Ekdawi 2021).

7.7.2.1 Morocco's "We Are Not Trash" Movement

A vehement environmental campaign was organized to protest the import of 2500 tons of waste from Italy in the wake of the Moroccan government's decision to receive 5.5% of Italy's annual waste (Karuri 2016). According to the 2016 agreement between the Moroccan and Italian governments, a total of 5 million tons of waste would be exported from Italy over a three-year period (Nasser 2016). The government claimed that the imported garbage was not toxic, and that it conformed to Article 42 of Law #28.00/2006 that regulates waste management and disposal in Morocco (Ngounou 2020). However public opinion, triggered by the media, environmental NGOs, and social networks, opposed this plan, arguing that the imported waste would have adversarial environmental and health impacts (Nasser 2016). The resulting movement managed to question the broader relationship between the State and national and transnational corporations within a market-based economy (Newell 2013). According to the environmental activists, the plan was to use this Italian waste to produce Refuse-Derived Fuel (RDF) as a cheap alternative to fossil fuel to power Morocco's cement industry (Karuri 2016; Reuters 2016). Activists argued that the State's ability to protect citizens' rights to a healthy environment was being compromised in order to provide favorable conditions for the greater profitability of private corporations.

This is in step with capitalism's anti-ecological configuration, whereby the State is primarily concerned with securing an optimal climate for private investment as a way of providing the tax revenues and employment opportunities necessary for buying public satisfaction and ensuring a regime's legitimacy. Hence corporate interests typically trump ecological priorities when the two are in conflict (Newell and Matthew 1998). The activists did not only highlight the State's role in advancing corporate interests but they also heavily criticized the role of corporations in blocking policy advances on other environmental issues such as climate action. At the time of the trash crisis, Morocco enacted Law #75-15 of July 2016, known as the "Zero Mika law", to prohibit the manufacture, import, export, commercialization and use of plastic bags (Climate Chance Observatory 2020). The "Zero Mika law" was part of nationwide efforts launched in conjunction with the 22nd Conference of the United Nations on Climate Change (COP22), hosted in Marrakech in 2016. It was seen as an indication of Morocco's concerted commitment to accelerate climate action (Bush 2016). In an interview, Hakima El Haite, the Minister of Environment at the time and an influential member of the Moroccan delegation at COP22, stated that importing the waste from Italy was necessary for the survival of Moroccan industry (Baker 2022). Here we have a situation where the actors who are charged with figuring out solutions for environmental problems (plastic pollution) are the same actors that maintain the conditions (cheap polluting energy sources) for corporate expansion and capital accumulation. This is a clear example of how climate change has become a post-political phenomena (Page 2022) where the responses to the climate crisis are articulated within the current capitalist system through the development of expensive technological fixes, tactical economic interventions and/or short-term environmental regulations that do not require far-reaching policy changes, do not disrupt the prevailing neoliberal economic structure and where the interests of corporations block the pursuit of rigorous climate action (Paterson 2021).

In addition, the Moroccan Coalition for Climate Justice (CMJC) managed to highlight the injustice of trade in solid waste between the Global North and the Global South, stressing that poorer countries should not be the garbage can of richer countries (Ngounou 2020). Reviews of the previous literature on environmental justice show that injustices that result from the global trade in waste have long been a key concern (Gregson et al. 2015). Distinct aspects of injustices are related to distribution of social, economic and environmental benefits and burdens between countries and actors. Also, injustices include the manner in which responsibility for causing environmental harms is allocated between Global Northern and Southern countries (Cotta 2020). For environmental justice scholars, the pursuit of high levels of industrialization to meet profligate consumption patterns of the Global North have resulted in massive amounts of waste that are often disposed of in the Global South (Thorpe 2021). Therefore, while consumers in the Global North maintain their luxurious lifestyles, their governments keep "trashing" the South by shipping polluting waste that exposes people in the Global South to various environmental and health risks (Gregson and Crang 2015). Environmental injustice also intersects with procedural justice; do the various actors in the society that receives the transboundary waste participate commensurably in the environmental decision-making (Bastos Lima and Gupta 2013)? By managing to convey the State's inability to secure citizens' rights to a healthy and safe environment for the sake of corporate profits, the "we are not trash" movement encouraged citizens to rethink State-society relations in a market-based economy, drawing attention to the inherent anti-ecological and environmental injustice aspects of the system. Activists continued to make the injustice evident through a variety of communication platforms accompanied with a public outcry and mobilization on the ground. The waste shipments were halted and the minister of the Environment was enforced to resign (Baker 2022).

7.8   Conclusion

The Arab region witnessed the eruption of numerous environmental movements that, regardless of their different mobilization practices, objectives and outcomes, all made use of the political dynamics and transformation that unfolded in the region during the brief political opening that immediately followed the overthrow of long-standing authoritarian regimes. These environmental movements not only reflect changing variables for activism but also subnational disparities and inequalities, characteristics of State-society relations and the adversaries of the prevailing market-based economic system in the region. One key variable that spurred the emergence of environmental movements was the increasing reliance on social media and other web-based tools (Moneer 2021). The spread of such tools helped environmental activists and other concerned citizens to get information about environmental violations from multiple sources, discuss their shared grievances, and reach out to sympathizers all over the region, spreading their messages, and amplifying their repertoire of protests (Clarke and Kocak 2020). In addition, the driving force behind these environmental movements did not come from what has conventionally been defined as formal organized civil society. Instead, the movements were triggered by and unfolded in the realm of informality—from unaffiliated citizens who belong to different social class and political backgrounds, which fostered the development of collective action and inclusive political participation (Gready and Robins 2017; Sika 2018).

These environmental movements were not a mere cry for a healthy environment (Moneer 2021). They underpinned deeper thinking about how nature and humans are interlinked in the everyday struggles of the masses in the Arab region to secure their environmental, political and economic rights. Beirut's 'You Stink' movement transformed from popular outrage about garbage piling up the streets into a popular movement against the corruption of Lebanon's sectarian political regime that proved incapable of providing basic services, from waste management to sanitation, health and education. Algeria's 'No to Fracking' movement provided a lens into rentier government, revealing the ways the government and international oil and gas corporations reproduce colonial relations through exploitation of cheap natural resources, imposing neocolonial structures of global trade and exacerbating pre-existing social and economic injustices faced by the vulnerable and marginalized people of the Sahara. Morocco's 'We Are Not Trash' movement did not only oppose the importing of trash from Italy for use as a cheap fuel in cement factories. Rather, it encouraged citizens to rethink State-society relations in a market-based economy, drawing attention to the inherent anti-ecological and environmental injustice aspects of the system.

References

Aarts P (1999) The middle east: a region without regionalism or the end of exceptionalism? Third World Q 20(5):911–925

Abaza M (2006) The changing consumer cultures of modern Egypt: Cairo urban reshaping. The American University in Cairo Press, Cairo

Abdelrahman M (2011) The transnational and the local: Egyptian activists and transnational protest networks. Br J Middle East Stud 38(3):407–424

Abu-Ismail K, Al-Kiswani B (2018) Extreme poverty in Arab states: a growing cause for concern. https://theforum.erf.org.eg/2018/10/16/extreme-poverty-arab-states-growing-cause-concern/

Abu-Ismail K, Hlasny V (2022) Wealth concentration rocketing in Arab countries after Covid-19. https://theforum.erf.org.eg/2022/03/23/wealth-concentration-rocketing-arab-countries-covid-19/

Abu-Rish Z (2017) Garbage politics. http://www.merip.org/mer/mer277/garbage-politics

Achcar G (2013) The people want. A radical exploration of the Arab uprising. Saqi, London

Adham K (2005) Globalization, neoliberalism, and new spaces of Capital in Cairo. Tradit Dwell Settl Rev 17(1):19–32

Afouxenidis A, Kourtelis C (2017) Reflections on neoliberal policy: a critical insight into recent development practices in Egypt and Morocco. Open J Pol Sci 7(2):291–310. https://doi.org/10.4236/ojps.2017.72024

Al Jeezrah (2020) Protesters in Tunisia halt key phosphate production. https://www.aljazeera.com/news/2020/11/25/protesters-in-tunisia-halt-key-phosphate-production

Al Rawashdeh R, Maxwell P (2013) Jordan, minerals extraction and the resource curse. Resour Policy 38(2):103–112. https://doi.org/10.1016/j.resourpol.2013.01.005

Alissa S (2007) The challenge of economic reform in the Arab world: toward more productive economies. Carnegie Endowment for International Peace, Washington, DC

Allal A, Bennafla K (2011) Les mouvements protestataires de Gafsa (Tunisie) et Sidi Ifni (Maroc) de 2005 à 2009. Revue Tiers Monde 5:27–45

Anguelovski I (2015) Tactical developments for achieving just and sustainable neighborhoods: the role of community-based coalitions and bottom-to-bottom networks in street, technical, and funder activism. Environ Plan C Gov Policy 33(4):703–725

Assouad L (2020) Inequality and its discontents in the Middle East. https://carnegie-mec. org/2020/03/12/inequality-and-its-discontents-in-middle-east-pub-81266

Atallah S (2015) Liberate the Municipal Fund from the Grip of Politicians. http://lcps-lebanon.org/ featuredArticle.php?id=52

Atwood B (2019) A city by the sea: uncovering Beirut's media waste. Commun Cult Crit 12:53–71. https://doi.org/10.1093/ccc/tcz011

Ayeb H (2011) Social and political geography of the Tunisian revolution: the alfa grass revolution. Rev Afr Polit Econ 38(129):467–479

Aziz A (2019) The South of Algeria has something to say. https://africasacountry.com/2019/07/the-south-has-something-to-say

Bahout J, Cammac P (2018) Arab political economy: pathways for equitable growth. https:// carnegieendowment.org/2018/10/09/arab-political-economy-pathways-for-equitable-growth-pub-77416

Baker L (2022) The sanitization of garbage politics: a case for studying waste at the local, state, and international politics in the MENA. https://pomeps.org/the-sanitization-of-garbage-politics-a-case-for-studying-waste-at-the-local-state-and-international-politics-in-the-mena

Bastos Lima MG, Gupta J (2013) The policy context of biofuels: a case of non-governance at the global level? Global Environ Polit 13(2):46–64

Bayat A (2013) The Arab spring and its surprises. Dev Chang 44(3):587–601

Beblawi H (1987) The rentier state in the Arab world. In: Beblawi H, Luciani G (eds) The rentier state. Routledge, London

Belakhdar N (2020) "Algeria is not for Sale!" Mobilizing against fracking in the Sahara. https:// merip.org/2020/10/algeria-is-not-for-sale-mobilizing-against-fracking-in-the-sahara/

Benkler Y (2006) The wealth of networks. Yale University Press, New Haven

Benli Altunışık M (2014) Rentier state theory and the Arab uprisings: an appraisal. Uluslararası İlişkiler 11(42):75–91. https://www.researchgate.net/publication/265178858_Rentier_State_ Theory_and_the_Arab_Uprisings_An_Appraisal

Berriane Y, Duboc M (2019) Allying beyond social divides: an introduction to contentious politics and coalitions in the Middle East and North Africa. Mediterr Polit 24:399–419. https://doi.org/ 10.1080/13629395.2019.1639022

Boersma T, Vandendriessche M, Leber A (2015) Shale gas in Algeria: no quick fix. Brookings Energy Security and Climate Initiative, Brookings Institute. https://www.brookings.edu/wp-content/uploads/2016/07/no_quick_fix_final-2.pdf

Borgen Project (2023) Five Facts about poverty in Algeria. https://borgenproject.org/

Brahimi A (2015) Algeria's cabinet reshuffle. Carnegie Endowment for International Peace. http:// carnegieendowment.org/sada/2015/06/02/algeria-s-cabinet-reshuffle/i9cn

Brechenmacher S (2017) Civil society under assault: repression and responses in Egypt, Ethiopia and Russia. Carnegie Endowment for International Peace, Washington, DC. https:// carnegieendowment.org/files/Civil_Society_Under_Assault_Final.pdf

Bullard RD, Johnson GS, Torres AO (2011) Environmental health and racial equity in the United States: building environmentally just, sustainable, and livable communities. American Public Health Association, Washington, DC

Bush D (2016) COP22 in Marrakech: A time for global action. https://medium.com/foggy-bottom/ cop22-in-marrakech-a-time-for-global-action-12eb8498d27a

Carothers T, Youngs R (2015) The complexities of global protests. https://carnegieendowment. org/2015/10/08/complexities-of-global-protests-pub-61537

Carty V (2015) Social movements and new technology. Westview, Boulder

Chaaban J (2016) One Year On, Lebanon's waste management policies still stink. https://www.lcps-lebanon.org/articles/details/1946/one-year-on-lebanon%E2%80%99s-waste-management-policies-still-stink

Chayes S (2018) Fighting the Hydra: lessons from worldwide protests against corruption. Carnegie Endowment for International Peace, Washington, DC

Chekir H, Diwan I (2013a) Distressed whales in the Nile_ Egypt capitislist in the wake of 2011 revolution. Working paper No. 747, Economic Research Forum. https://erf.org.eg/app/uploads/2014/03/747.pdf

Checkir H, Diwan I (2013b) Crony capitalism in Egypt. CID workin papers No. 250. https://www.hks.harvard.edu/sites/default/files/centers/cid/files/publications/faculty-working-papers/250_Diwan_EGX+paper.pdf

Civil Society Knowledge Center (2014) Waste Management Conflict. https://civilsociety-centre.org/timeliness/4923#event-a-href-sir-sit-naameh-landfill-suspended-after-crackdownsit-in-at-naameh-landfill-suspended-after-crackdown-a

Civil Society Knowledge Center (2016)Waste Management Conflict. Retrieved from https://civilsociety-centre.org/timeliness/4923#event-a-href-sir-sit-naameh-landfill-suspended-after-crackdownsit-in-at-naameh-landfill-suspended-after-crackdown-a

Clarke K, Kocak K (2020) Launching revolution: social media and the Egyptian uprising's first movers. Br J Polit Sci 50:1025–1045

Climate Change Observatory (2020) Moroccan society's uneven response to the proliferation of waste. https://www.climate-chance.org/wp-content/uploads/2020/03/cp-waste_morocco_english.pdf

Cooley A (2001) Booms and busts: theorizing institutional formation and change in oil states. Rev Int Polit Econ 8(1):167

Cotta B (2020) What goes around, comes around? Access and allocation problems in Global North–South waste trade. Int Environ Agreements 20:255–269. https://doi.org/10.1007/s10784-020-09479-3

Crapanzano V (2010) The wound that never heals. Alif: journal of comparative. Poetics 30:57–84

Davydova A (2021) Environmental Activism in Russia: Strategies and prospects. https://www.csis.org/analysis/environmental-activism-russia-strategies-and-prospects

Dejoui N (2018) Résultats des élections municipales 2018. https://www.leconomistemaghrebin.com/2018/05/09/resultat-final-de-lisie/

Diwan I, Keefer Ph, Schiffbaur M (2013) The effect of cronyism on private sector growth in Egypt. https://www.femise.org/wp-content/uploads/2015/10/Diwan.pdf

Dwivedi R (1998) Environmental movements in the global south: issues of livelihood and beyond. Int Sociol J Int Sociol Assoc 16(1):11–31

Eder K, Donati P, Dinai M, Statham P, Szerszynski B, Strzdom P, Drom D, Le Saout D, Poferl A, Brand K, Ruzza C, Ibarra P (1995) Framing and communicating environmental issues. Final Report to the Commission of the European Communities

Ekdawi A (2021) Beirut's "You Stink" movement: a tongue in cheek slogan to hold officials accountable. https://accountabilityresearch.org/beiruts-you-stink-movement-a-tongue-in-cheek-slogan-to-hold-officials-accountable/

El Erian M, Tareq S (1993) Economic reform in Arab countries: a review of structural issues for the remainder of the 1990s. International Monetary Fund. https://doi.org/10.5089/9781451845754.001

El Ksayer LN (2017) Can co-production partnerships lead to new forms of sustainable urban governance? Waste management under conditions of enduring political crisis in Beirut. https://www.urbanmanagement.tuberlin.de/fileadmin/f6_urbanmanagement/Study_Course/2017_Thesis_Lea_Nassif_El_Ksayer.pdf

Elgendy K, Abaza N (2020) Urbanization in the MENA region: a benefit or a Curse? https://mena.fes.de/press/e/urbanization-in-the-mena-region-a-benefit-or-a-curse

Elliott L (2017) Environmentalism. https://www.britannica.com/topic/environmentalism

El-Mikawy N (2020) From risk to opportunity: Local governance in the southern Mediterranean. https://www.cidob.org/en/publications/publication_series/notes_internacionals/n1_232/from_risk_to_opportunity_local_governance_in_the_southern_mediterranean

Escribano G (2018) Algerian presidential elections and the energy reform agenda. https://www.realinstitutoelcano.org/en/analyses/algerian-presidential-elections-and-the-energy-reform-agenda/

Fahmy N (2002) The politics of Egypt: state-society relationship. Routledge, London/New York

Farid S (2019) Can Egypt's slums be replaced by 'Safe Zones'? https://english.alarabiya.net/features/2019/03/31/Can-Egypt-s-slums-be-replaced-by-safe-zones-

Farouk M (2020) Cautious hopes for slum dwellers relocated in Egypt housing project. https://www.reuters.com/article/us-egypt-cities-housing-feature-trfn-idUSKCN25A0DP

Finaz C (2015) A new climate for peace: Egypt country risk brief, Briefing note #9. International Alert, London. Retrieved from: https://library.ecc-platform.org/publications/egypt-climate-fragility-risk-brief

Gray M (2011) A theory of "late Rentierism" in the Arab states of the Gulf. Center for International and Regional Studies. https://www.files.ethz.ch/isn/134326/No_7_MatthewGrayTheoryLaterRentierismArabStatesGulf.pdf

Gready P, Robins S (2017) Rethinking civil society and transitional justice: lessons from social movements and 'new' civil society. Int J Hum Rights 21(7):956–975. https://doi.org/10.1080/13642987.2017.1313237

Green F (2018) Anti-fossil fuel norms. Clim Chang 150:103–116. https://doi.org/10.1007/s10584-017-2134-6

Gregson N, Crang M (2015) From waste to resource: the trade in wastes and global recycling economies. Rev Environ Res 40:151–176

Gregson N, Crang M, Fuller S, Holmes H (2015) Interrogating the circular economy: the moral economy of resource recovery in the EU. Econ Soc 44(2):218–243

Gylfason T (2001) Natural resources, education, and economic development. Eur Econ Rev 45(4–6):847–859

Hameed S (2020) Political economy of Rentierism in the Middle East and disruptions from the digital space. Contemp Rev Middle East 7(1):54–89. https://doi.org/10.1177/2347798919889782

Hamouchene H (2020) Energy transitions and colonialism. https://www.cadtm.org/spip.php?page=imprimer&id_article=19123

Hamouchene H (2022) The energy transition in North Africa: Neocolonialism again. https://longreads.tni.org/ar/the-energy-transition-in-north-africa-neocolonialism-again

Hamouchene H, Pérez A (2016) Energy colonialism: the EU's gas grab in Algeria. The Observatory on Debt and Globalisation (ODG). https://odg.cat/wp-content/uploads/2017/06/energy_colonialism_algeria_eng_0.pdf

Hanieh A (2015) Crisis, conflict and the political economy of the Middle East Region. https://www.iemed.org/publication/crisis-conflict-and-the-political-economy-of-the-middle-east-region/

Harders C (2015) "State analysis from below" and political dynamics in Egypt after 2011. Int J Middle East Stud 47(1):148–151. https://doi.org/10.1017/S0020743814001524

Hendawy M, Saeed A (2019) Beauty and the beast: the ordinary city versus the mediatized city: the case of Cairo. Urbanization 4(2):126–134

Henry M, Springborg R (2010) Globalization and the politics of development in the Middle East. 2nd Edition, Cambridge University Press, Cambridge

Herb M (2005) No representation without taxation? Rents, development, and democracy. Comp Polit 37(3):297–316

Hertog S (2023) Insiders and outsiders: the political economy of Arab labour markets. https://theforum.erf.org.eg/2023/07/11/insiders-and-outsiders-the-political-economy-of-arab-labour-markets/

Heydemann S (2016) Explaining the Arab uprisings: transformations in comparative perspective. Mediterr Polit 21(1):192–204. https://doi.org/10.1080/13629395.2015.1081450

Human Rights Watch (UNHCR) (2007) Monopolizing Power: Egypt's Political Parties Law. Retrieved from: https://www.refworld.org/reference/countryrep/hrw/2007/en/38506? prevDestination=search&prevPath=/search?keywords=egypt+political+parties&order= desc&sm_document_source_name=Human+Rights+Watch&sort=score&result=result-3850 6-en

Human Rights Watch (2020) Lebanon: huge cost of inaction in trash crisis

Hyde M (2011) Is Egypt ready for a New Green political party? https://www.egyptindependent. com/egypt-ready-new-green-political-party/

International Crisis Group (2016) Algeria's South: Trouble's Bellwether. https://www.crisisgroup. org/middle-east-north-africa/north-africa/algeria/algeria-s-south-trouble-s-bellwether

International Labor Organization (ILO) (2012) Rethinking economic growth: Towards Productive and Inclusive Arab Societies. https://www.ilo.org/beirut/publications/WCMS_208346/lang%2 D%2Den/index.htm

Issa P, Zaatari M (2015) Officials wrangle over Beirut and Mount Lebanon's waste. The Daily Star (Lebanon). Retrieved from https://www.pressreader.com/lebanon/the-daily-star-lebanon/20150 724/281556584521792

Jacobs R (2013) Nuclear Conquistadors: Military Colonialism in Nuclear Test Site Selection during the Cold War. Asian J Peacebuilding 1(2):157–177

Jenkins K (2018) Setting energy justice apart from the crowd: lessons from environmental and climate justice. Energy Res Soc Sci 39:117–121

Karduni A, Sauda E (2020) Anatomy of a protest: spatial information, social media, and urban space. Soc Media Soc 1(15). https://doi.org/10.1177/2056305119897320

Karuri K (2016) Morocco bans importing waste from Italy after uproar. https://www.africanews. com/2016/07/16/morocco-bans-importing-waste-from-italy-after-uproar/

Keck M, Sikkink K (2014) Activists beyond Borders: advocacy networks in international politics. Cornell University Press, Ithaca. https://doi.org/10.7591/9780801471292

Kerbage C (2017) Politics of coincidence: the Harak confronts its peoples. Retrieved from https:// website.aub.edu.lb/ifi/publications/Documents/working_papers/20170213_wp_hirak_ english.pdf

Kettell S (2020) Oil crisis. Encyclopedia Britannica. https://www.britannica.com/topic/oil-crisis. Accessed 24 Aug 2021

Khouri R (2019) How poverty and inequality Are Devastating the Middle East. https://www. carnegie.org/topics/topic-articles/arab-region-transitions/why-mass-poverty-so-dangerous-mid dle-east/

Kingston P (2019) The Ebbing and flowing of political opportunity structures: revolution, counterrevolution, and the Arab uprisings. In: Arce M, Rice R (eds) Protest and democracy. University of Calgary Press, Calgary. http://hdl.handle.net/1880/110581book

Kinniburgh C (2015) The Fracktivists. https://www.dissentmagazine.org/article/anti-fracking-new-york-global-algeria-poland/

Kirsanli F (2023) Crony Capitalism and Corruption in the Middle East and North Africa. Journal of Economy Culture and Society. https://doi.org/10.26650/JECS2023-1210965

Kostovicova D, Bojičić-Dželilović V (2013) Introduction: civil society and multiple transitions: meanings, actors and effects. In: Bojičić-Dželilović V, Ker-Lindsay J, Kostovicova D (eds) Civil society and transitions in the Western Balkans. Palgrave Macmillan, London, pp 1–28

Lacouture M (2021) Privatizing the commons: protest and the moral economy of national resources in Jordan. IRSH 66:113–137

Lageman T (2016) The Crew Bringing Recycling to Tunis. https://www.bloomberg.com/news/ articles/2016-09-20/tunis-recyclage-collects-recycling

Loewe M, Albrecht H (2022) The social contract in Egypt, Lebanon and Tunisia: what do the people want? J Int Dev 35:838–855. https://doi.org/10.1002/jid.3709

Longeray P (2015) Violence Flares over Halliburton's fracking tests in Algeria. https://www.vice. com/en/article/qva75q/violence-flares-over-halliburtons-fracking-tests-in-algeria

Mahdavy H (1970) The patterns and problems of economic development in rentier states: the case of Iran. In: Cook MA (ed) Studies in the economic history of the Middle East. School of Oriental African Studies/Oxford University Press, London, pp 428–467

Malik A (2019) A pyramid of privilege: crony capitalism in the Middle East. https://www.qeh.ox.ac.uk/blog/pyramid-privilege-crony-capitalism-middle-east

Mihalyi P, Szelenyi I (2019) Rent seeker profits and wages and inequalities the top 20%. Palgrave Macmillan, Singapore

Mills R, Alhashemi F (2018) Resource regionalism in the Middle East and North Africa: rich lands, neglected people. Brookings Doha Center

Mohai P, Saha R (2015) Which came first, people or pollution? A review of theory and evidence from longitudinal environmental justice studies. Environ Res Lett 10:125011. https://doi.org/10.1088/1748-9326/10/12/125011

Moneer Z (2021) Environmental movements after the Arab spring. In Environemntal Poltics in the Middle East and North Africa: proceedings from first inaugural conference. Arab Reform Initiative https://s3.eu-central-1.amazonaws.com/storage.arab-reform.net/ari/2021/09/2317320 5/ENGLISH-Environmental-politics-in-the-Middle-East-and-North-Africa-Proceedings-from-First-Inaugural-Conference-final.pdf

Moneer Z (2022) Moving beyond climate coloniality. https://www.mei.edu/publications/moving-beyond-climate-coloniality

Movahed M (2016) Does capitalism have to be bad for the environment? https://www.weforum.org/agenda/2016/02/does-capitalism-have-to-be-bad-for-the-environment/

Murphy E (1998) Legitimacy and economic reform in the Arab world. J North Afr Stud 3:71–92. https://doi.org/10.1080/13629389808718338

Naggar M (2020) Infrastructure of refugee settlements in Lebanon reflects their social status. https://www.dandc.eu/en/article/infrastructure-also-question-integration-can-be-seen-refugee-camps-lebanon

Nasser Z (2016) Importation of Toxic Italian garbage to Morocco sparks outrage. . . "greenarea.me" receives legal analysis from Basel Action Network. http://greenarea.me/en/159202/importation-toxic-italian-garbage-morocco-sparks-outrage-greenarea-receives-legal-analysis-basel-action-network

Newell P (2013) Globalization and the environment: capitalism, ecology and power. Wiley

Newell P, Matthew P (1998) A climate for business: global warming, the state and capital. Rev Int Polit Econ 5(4):679–704

Ngounou B (2020) Morocco: controversies over import of combustible waste. https://www.afrik21.africa/en/morocco-controverses-over-import-of-combustible-waste/

Ottaway M (2021a). The Forgotten Link: Political Parties and Political Reform in the Arab World. Retrieved from: https://www.wilsoncenter.org/article/forgotten-link-political-parties-and-politi cal-reform-arab-world

Ottaway M (2021b). Abdicating Responsibility: Political Parties in Egypt. Retrieved from: https://www.wilsoncenter.org/article/abdicating-responsibility-political-parties-egypt

Owen R (2015) The history of the Middle East in the postcolonial period. Commentary Ortadoğu Etütleri 7(1):9–15

Oxford Poverty and Human Development Initiative (OPHI) (2018) Global multidimensional poverty index 2018: the most detailed picture to date of the world's poorest people. Report. University of Oxford Press, ISBN 978–1–912291-12-0

Özekin M, Arıöz Z (2014) Rethinking the conflict-proneness of oil-rentier states in historical context. Altern Turk J Int Relat 13(1–2). https://dergipark.org.tr/tr/download/article-file/19336

Page J (2022) The environmental humanities: responding to the "post-political" phenomenon of climate change. https://talkinghumanities.blogs.sas.ac.uk/2022/06/10/the-environmental-humanities-responding-to-the-post-political-phenomenon-of-climate-change/

Patel D (2018) How to lose momentum in five steps: why did Lebanon's You Stink movement fail? https://blogs.lse.ac.uk/internationaldevelopment/2018/11/30/how-to-lose-momentum-in-five-steps-why-did-lebanons-you-stink-movement-fail/

Paterson M (2021) Climate politics between conflict and complexity. In: Böhm S, Sullivan S (eds) Negotiating climate change in crisis. Open Book Publishers, Cambridge. https://doi.org/10. 11647/OBP.0265

Patrick S (2012) Why natural resources are a curse on developing countries and how to fix it. https://www.theatlantic.com/international/archive/2012/04/why-natural-resources-are-a-curse-on-developing-countries-and-how-to-fix-it/256508/

Plaetzer N (2014) Protests and revolutions in the 21st century. J Int Aff 68(1):255–265

Pope C (2016) Tunisia's thirst uprising: a nation on edge. Circle of Blue. http://www.circleofblue.org/2016/africa/a-nation-on-the-edge/

Puranen B, Widenfalk O (2007) The rentier state: does Rentierism hinder democracy? In: Moaddel M (ed) Values and perceptions of the Islamic and middle eastern publics. Palgrave Macmillan, New York. https://doi.org/10.1057/9780230603332_7

Reuters (2016) Environmental protests spur Morocco to halt waste imports for energy. https://www.reuters.com/article/us-morocco-environment/environmental-protests-spur-morocco-to-halt-waste-imports-for-energy-idUKKCN0ZU26R

Roberts J (1986) The effects of the oil price collapse on the Gulf cooperation council economics. J Energy Dev 12(1):103–114

Rodríguez-Labajos B, Yánez I, Bond P, Greyl L, Munguti S, Uyi Ojo G, Overbeek W (2019) Not so natural an Alliance? Degrowth and environmental justice movements in the global south. Ecol Econ 157:175–184. https://doi.org/10.1016/j.ecolecon.2018.11.007

Rowlands L and N Peña (2019) We will not be silenced. Climate activism from the frontlines to the UN. https://www.civicus.org/documents/WeWillNotBeSilenced_eng_Nov19.pdf

Rubiano M (2021) 2020 was the deadliest year for environmental activists. Here's why. https://grist.org/protest/2020-was-the-deadliest-year-for-environmental-activists-heres-why/

Rutledge E (2013) The Rentier State/Resource Curse narrative and the state of the Arabian Gulf. https://mpra.ub.uni-muenchen.de/59501/1/MPRA_paper_59501.pdf

Sadiki L (2019) Regional development in Tunisia: the consequences of multiple marginalizations. https://www.brookings.edu/research/regional-development-in-tunisia-the-consequences-of-multiple-marginalization/

Sadiki L, Wimmen H, Al-Zubaidi L (2013) Democratic transition in the Middle East: unmaking power. Routledge, London

Salanova R (2012) Social media and political change: the case of the 2011 revolutions in Tunisia and Egypt. Institut Català Internacional per la Pau, Barcelona

Shalby M (2016) Taking political parties seriously in the Arab world. Edward P. Djerejian Center for the Middle East. https://www.bakerinstitute.org/research/taking-political-parties-seriously-arab-world

Sharp D (2018) The urbanization of power and the struggle for the City. Middle East Research and Information Project: Critical Coverage of the Middle East Since 1971. https://merip.org/2018/10/the-urbanization-of-power-and-the-struggle-for-the-city/

Shechter R (2008) The cultural economy of development in Egypt: economic nationalism, hidden economy and the emergence of mass consumer society during Sadat's Infitah. Middle East Stud 44(4):571–583

Sika N (2012) The political economy of Arab uprisings. The European Institute of the Mediterranean, Girona

Sika N (2018) Civil society and the rise of unconventional modes of youth participation in the MENA. Middle East Law Gov 10:237–263

Smiley K, Emerson M (2020) A spirit of urban capitalism: market cities, people cities, and cultural justifications. Urban Res Pract 13(3):330–347. https://doi.org/10.1080/17535069.2018.1559351

Sowers J (2012) Environmental politics in Egypt: activists, experts and the state (1). Routledge https://doi.org/10.4324/9780203808979

Sowers J (2018) Environmental activism in the Middle East and North Africa. In: Verhoven H (ed) Environmental politics in the Middle East: local struggles, global connections. C. Hurst & Co. Ltd., London, pp 1–27

State Information Service (SIS) 2023 Political Parties in Egypt. https://www.sis.gov.eg/section/10/2 58?lang=en-us

Strobl J (2016) Social movements challenging environmental politics? Reframing the garbage crisis in Lebanon. Retrieved from https://scholarworks.aub.edu.lb/bitstream/handle/10938/11020/t-64 90.pdf?sequence=1&isAllowed=y

Tadamun (2016) The Garbage Crisis in Lebanon: From Protest to Movement to Municipal Elections. http://www.tadamun.co/garbage-crisis-lebanon-protest-movement-municipal-elec tions/?lang=en

Taylor D (2000) The rise of the environmental justice paradigm: injustice framing and the social construction of environmental discourses. Am Behav Sci 43(4):508–580

Thorpe D (2021) The global waste trade has created "sacrifice zones" for health and the environment. https://thefifthestate.com.au/columns/spinifex/the-global-waste-trade-has-created-sacri fice-zones-for-health-and-the-environment/

Tolba MK, Saab NW (eds) (2008) Arab environment: future Challenges. Arab Forum for Environment and Development (AFED), Technical Publications and Environment and Development Magazine, Beirut, 7–8

Toledo H (2013) The political economy of emiratization in the UAE. J Econ Stud 40:39–53. https://doi.org/10.1108/01443581311283493

Tullock G (1989) The economics of special privilege and rent seeking. Springer, New York City

UN Habitat (2003) The challenge of slums: global report on human settlements. https://unhabitat.org/sites/default/files/download-manager-files/The%20Challenge%20of%20Slums%20-%20 Global%20Report%20on%20Human%20Settlements%202003.pdf

Urkidi L, Walter M (2011) Dimensions of environmental justice in anti-gold mining movements in Latin America. Geoforum 42(6):683–695

Werenfels I (2016) Fracking in Algeria on Twitter: connecting the Periphery to the Center and the World, #InSalah. SWP Research Paper 5

White R (2013) Transnational environmental crime. Routledge, New York

Williams C (2013) Strategy and tactics in the environmental movement. https://climateandcapitalism.com/2013/09/21/strategy-tactics-environmental-movement/

Wood J (2015) Beirut's trash war pushes Lebanon to the brink. Retrieved from https://www.thenational.ae/world/beirut-s-trash-war-pushes-lebanon-to-the-brink-1.26502

World Bank (2021a) Demographic trends and urbanization. https://www.worldbank.org/en/topic/urbandevelopment/publication/demographic-trends-and-urbanization

World Bank (2021b) Urban population-Middle East & North Africa. https://data.worldbank.org/indicator/SP.URB.TOTL.IN.ZS?locations=ZQ

Yacoubian M (2018) Lebanon's new election law results in limited change. Retrieved fromhttps://www.usip.org/blog/2018/05/lebanons-new-election-law-results-limited-change?fbclid= IwAR0lbvnhyRazPkUAAHo5XPmFcXd0p3zNHmFxwyTBZsJHdjTjO9MxC-ce35Y

Yom S (2015) Arab civil society after the Arab spring: weaker but deeper. https://www.mei.edu/publications/arab-civil-society-after-arab-spring-weaker-deeper

York J (2010) Morocco: a new green party https://globalvoices.org/2009/04/14/morocco-a-new-green-party/

Yoshida M (2016) Capacity development in environmental management administration through raising public awareness: a case study in Algeria. JICA Research Institute. https://www.jica.go.jp/jica-ri/publication/workingpaper/wp_176.html

Part II
Urban Planning and Development

Chapter 8
Power and Planning Discourse: Defining the Greater Amman Metropolitan Area

Fuad Malkawi

8.1 Introduction

When planners are asked about their work, their answer is usually simple: Planners make plans. They usually describe their work as processes that entail solving preexisting problems. Indeed, they often champion themselves as the *equipped experts* who can diagnose planning problems and the only party who can provide the solutions. Problems and solutions are usually portrayed in planning discourse as technical issues that are removed from any societal or historical event. While the planners recognize the political and societal pressure, they must face when making plans, they still distance their plans from any hidden agenda, whether it be social or political. Their acts are rather introduced as innovative, original acts that respond only to technical issues.

As a result, planning is presented as a purely technical field. This description of the planners' work may apply in certain cases. There are occasions, however, in which planning entails more than just the technical issues. Many planning acts are often repeated and accepted, and they become practices in the historical sense, the sense that goes beyond individuals. Such practices are usually de-historicized and naturalized to hide either a political agenda or an ideological commitment. Presented as solutions to technical problems, historical practices often precede the problems they are supposed to solve. Accordingly, planners in such cases construct problems rather than diagnose them, to legitimize those historical practices through presenting them as solutions.

F. Malkawi (✉)
Senior Urban Specialist at the World Bank, Washington, DC, USA
e-mail: fmalkawi@worldbank.org

© The Author(s), under exclusive license to Springer Nature Switzerland AG 2024
K. Darmame, E. Ross (eds.), *Local Governance and Development in Africa and the Middle East*, Local and Urban Governance,
https://doi.org/10.1007/978-3-031-60657-1_8

A good case in point is the making of Greater Amman (GA) in Jordan's capital. In 1985, GA was pronounced a city-region with a metropolitan authority, as a preliminary step in the process of making the Greater Amman Comprehensive Development Plan (GACDP). The planners presented the idea of GA as an innovative, original idea that had no historical precedents in Jordan. The acts that produced it and defined its borders, size, and form of governance were presented in the plan as solutions to preexisting technical problems. However, the planners' representation of the idea gives only part of the whole picture. The framing idea of GA was rather delimited by social and political practices; such was the case with many planning acts. Within the latitude that the planners had, they were disciplined by their own ideological commitments. Moreover, the idea of GA is a practice in the historical sense. It has been often repeated and accepted as a form of city governance for Amman since the sixties till it was fully adopted in the mid-eighties as part of GACDP. However, the idea was dehistoricized and naturalized. The two processes of dehistoricization and naturalization, as will be explained later, hided the planners' quest for control over planning the city and its surroundings. Perhaps this issue of control, which the GA planners view as a crucial ingredient of good planning, is what led to the making of GA in its current form. Regardless of its technical rationale, the idea of a metropolitan authority in Jordan's capital meant providing the planners with control over the whole area of GA while denying it to other parties, namely the local authorities.

8.2   The Idea of Greater Amman

In 1983, Jordan received a $30 million loan from the World Bank to implement a package of improvements in infrastructure, garbage collection, land use planning, and management and engineering capacities in its capital, Amman. The planners of Amman Municipality decided that these measures of improvements should be incorporated in a long-term plan for the whole GA area. They argued that the main problem of the city was the absence of a comprehensive plan to which all local authorities surrounding Amman should abide. Moreover, the planners argued that a metropolitan authority should be created, as a first step in the making of the plan, to ensure the plan would be implemented over the whole area. The mayor of Amman at the time, Abd al-Ra'uf al-Rawabidah, adopted the idea and was able to get it approved by the cabinet of Jordan. In March 1985, he wrote to the Minister of Municipal and Rural Affairs and the Environment (MMRAE), asking him to suspend all planning activities within the area defined as GA. The various municipalities and village councils within GA were amalgamated into one metropolitan authority— the Greater Amman Municipality—and the preparation of the GACDP was launched.

Preparing a comprehensive plan was too big of a job for the inexperienced local planners of Amman municipality. Hence, a joint team was created with planners from Amman Municipality and Dar al-Handasah Consultants, a private international firm. The idea behind creating this team (hereafter, the Joint Team) was to mix international expertise with local knowledge. The Joint Team was headed by John Calder, a British planner from Dar al-Handasah. Calder had many chief planners working under him during the five-year period of the plan. Experts on transportation, infrastructure, demography, and economics were brought in on a short-term basis. Calder was joined toward the end by Husni Iskandar, a senior planner from Dar al-Handasah, whose job was to negotiate with city officials on certain legal aspects delaying the adoption of the plan.

The GACDP was completed in 1988 (almost two years after the anticipated completion date) and was adopted in 1990 as the first comprehensive plan for GA. In addition to land-use planning, transportation, and the urban design issues, GACDP provided an in-depth appraisal of management capabilities of the various administrations within the municipality. The plan built on several of the small-scale plans that preceded it, but its main source of information was the Five-Year Economic and Social Plan 1986–1990. The GACDP was never implemented, though. The Master Plan, presented as the heart of the GACDP, was obsolete by the time the comprehensive plan came out in its final form. Following its adoption in 1988, the GACDP functioned as policy guidelines for Greater Amman Municipality, which in technical terms means, according to one planner, that it has been "put on the shelf." Nevertheless, the GACDP was important because it established the centrality of the idea of "Greater Amman."

Serious thinking about defining GA emerged several years before the comprehensive plan, though. In 1980, a group within the MMRAE presented its vision of how the Greater Amman Region (GAR) should look in the ministry's magazine *Baladi*.[1] In the group's vision, GAR included two additional main cities—Zarqa and Ruseifa—as well as a group of small municipalities and village councils. These two cities depended heavily on Amman, the ministry group argued, since the majority of their populations worked there. Later, the group excluded Zarqa and Ruseifa on the ground that the two cities were proven to be independent of Amman (Greater Amman Municipality 1985, 3). In the end, the group's ideas were used by the planners of the Joint Team, who built on their ideas of GAR in defining a study area for the comprehensive plan. However, unlike the ministry group, the Joint Team planners argued that "Greater Amman should be as compact as possible" (ibid., 5). They set criteria for inclusion, which included "interdependency, accessibility, and

---

[1]The debate which took place on the pages of *Baladi*, a monthly magazine published by the MMRAE, presented the ministry's new ideas about the boundaries of GA and the kind of administration it should have. No copies of that issue of the magazine could be found, but the vision was narrated by many of the officials interviewed for this study, and the story of GAR was fully explained in the GACDP's Boundaries Report as an entry point to defining the GA area. GA was to consist of 19 municipalities with a population of almost one million and expand over an area of over than 230 square miles. Zarqa and Ruseifa were excluded from this area. The question of how to understand the metropolises and/or city-regions would remain a relevant question in Jordan and the whole Middle East for years to come. See Malkawi (2008).

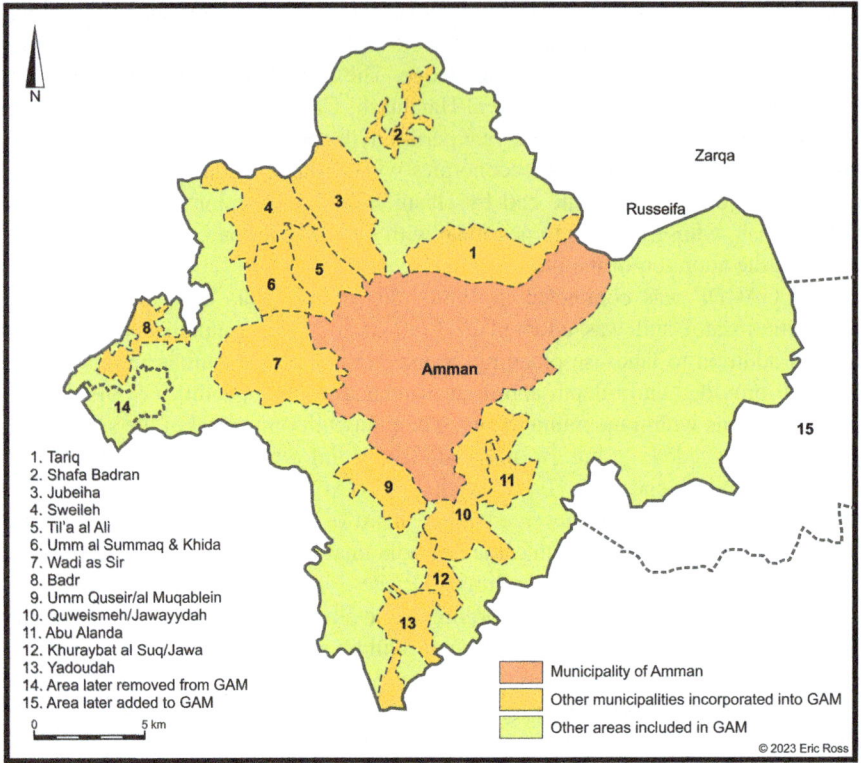

N

Zarqa

Russeifa

4   3

2

1

6   5

8

7   Amman

14

15

1. Tariq
2. Shafa Badran
3. Jubeiha
4. Sweileh
5. Til'a al Ali
6. Umm al Summaq & Khida
7. Wadi as Sir
8. Badr
9. Umm Quseir/al Muqablein
10. Quweismeh/Jawayydah
11. Abu Alanda
12. Khuraybat al Suq/Jawa
13. Yadoudah
14. Area later removed from GAM
15. Area later added to GAM

9   11

10

12

13

0   5 km

Municipality of Amman

Other municipalities incorporated into GAM

Other areas included in GAM

© 2023 Eric Ross

Fig. 8.1 Municipalities of Greater Amman prior to amalgamation

the pattern and intensity of development" in the various administrative units surrounding Amman (ibid.). These criteria were not defined clearly. Nonetheless, the planners cited them as the basis for their decision to exclude many of the administrative units included in the MMRAE group's definition of GA.[2] More than 25 municipalities and village councils were included in the Joint Team's definition of GA (see Fig. 8.1).

Though the issue of boundaries was a major concern for the planners of the Joint Team, they were more concerned with the administration of GA. The metropolis's growth and development needed to be tamed and controlled, the team members argued. Such a task, according to them, could not be achieved under the current "multiplicity of administrative units" (Municipality of Amman 1984, 2). Indeed, the

[2]Despite these declared criteria, the planners of the team had to follow existing municipal boundaries in defining the boundaries of GA. The confusion which might have been caused had they included parts of a municipality discouraged them from such act. Thus, many of the exclusion decisions were not really based on the stated technical issues.

multiplicity of administrative units was probably *the* major problem in the eyes of the planners since the creation of GA. For the planners, it was important to "pursue a policy of securing the co-operation of all the authorities in recognizing that the objectives of the plan and any necessary changes were for the benefit of the whole community of Greater Amman" (ibid., 2). The planners of the Joint Team envisioned a *two-tier system* as a form of administration that could be applied in GA, as follows:

- A first-tier authority (metropolitan authority) would exercise major strategic functions, including responsibility for the control, monitoring, review and implementation of the Master Plan over the whole of Greater Amman. This authority would also be responsible for dealing with all major planning/building applications and departures from the strategic plan and would establish a new consultative system for dealing with development control at both the first and second tier levels (Greater Amman Municipality 1985, 22).
- At the second-tier level, municipalities would be combined to form larger units. Those would be responsible for dealing with planning and building licenses in accordance with consultation systems that are supervised by the first-tier authority. Local governments would also be responsible for initiating local plans, which of course would have to be approved by the first-tier government. As the team wanted it, the central—metropolitan—authority would have *power over* the whole of GA. Indeed, the proposed *code of practice* described in the team's Boundaries Report[3] ties almost every planning, or architectural, issue the local government deals with to the central authority in one way or another.

The concept of a central, or metropolitan, governmental body was not new to the region. The GAR has always been under the jurisdiction of more than one council, each functioning as a metropolitan authority for a period (namely, the Amman Urban and Regional Council and the Amman Urban Group). However, these councils were always attached to the MMRAE, which reduced the councils' effectiveness and limited their power. What the planners of the Joint Team were really after was an increase and a shift in power over the GAR. The planners were granted this in March of 1985, when the mayor of Amman, al-Rawabidah, accepted the idea of GA and took it to the National Assembly, which in turn approved the idea. Shortly after, all the municipalities and village councils in the area were amalgamated into *Greater Amman Municipality* (GAM). This was the first step toward the making of the GACDP. It proposed restructuring the local jurisdictions to form larger areas, with the various local governments within GA's boundaries directly attached to the central authority (Fig. 8.1).

---

[3] The Boundaries Report is one of the reports the team prepared as a preparatory step toward writing the plan (Greater Amman Municipality 1985).

8.3   The Idea of GA as an Historical Practice

For something to be characterized as a practice in the historical sense, the sense that goes beyond individuals, there must be some sort of sameness or identity between the practice at different stages (Turner 1994, 19).[4]

The history of planning in Amman indicates that the planners of the Joint Team were not inventing a new idea when they defined GA. Indeed, the idea of GA had been introduced almost two decades before the creation of the Joint Team. And, as explained in this chapter, it has been reintroduced many times afterward. However, the Joint Team planners referred to the historical background of the idea only in a passing manner. They did so in order to explain their choices of administrative units that were to be included in GA. They didn't make any reference to the idea's origin or the history of its development. In that sense, they dehistoricized the idea, making it appear as the natural solution to a preexisting technical problem.

The idea of GA came to Jordan through the British planners who worked on planning assignments or held planning positions in Amman in the early 1960s. Vernon Newcombe[5] was one of those planners. In an article published in 1967, he explained the line of thinking he and his colleagues had while working on a planning assignment for the United Nations in Amman. It seemed important, he wrote, "to make planning independent of accidental local authority boundaries" (Newcombe 1967, 415). Consequently, the planners of that period pushed toward the designation of a metropolitan planning area. Newcombe, and his colleagues, wanted a central metropolitan authority that could be responsible for all planning aspects of the city. Soon, this authority was recognized, and the planning area was designated.

The British planners were learning from the London experience what could be done in Amman. While no actual text can confirm this link, it is suggested by the similarities between the various planning concepts—from the idea of *city-region*, all the way to the *two-tier system*—that were implemented in the two cities. For example, it was accepted among British planners of the late 1950s and early 60s that a *city-region* would be the right basis for local government reform. In 1963, a metropolitan authority for Greater London was created. The act that created the Greater London Council came as a reform of London local government (Hall 1975,

---

[4]For Turner, the term "practice" refers to the repetition of certain acts in time and space. While some views describe practices as causes and others as presuppositions, he argues, none of these views can explain the transmission of the sameness in time and space. Despite his struggle to formulate a theory that explains how practice is transmitted without being habituated or governed by rules (as Wittgenstein would have it), Turner seems to have imposed this idea of sameness on practice. Admitting this imposition, he uses the concept of historical sameness, rather than repetition, to state that practices are, in Bourdieu's words, *produced and reproduced*.

[5]Newcombe, a British planner, worked in Amman on a town-and-country planning assignment sponsored by the United Nations in the early 1960s. His major assignment was to participate in the committee that was drafting new legislation. Nevertheless, he played a role in many other planning activities. See Newcombe (1964) and (1967)

174).[6] It was later argued that the changing situation of London in the 1960s required *a new two-tier system of plan-making.*[7] This argument was accepted and embodied in a Planning Act in 1968. Similarly, in 1963 the Jordanian Council of Ministries approved the creation of the GAR. Since then, the idea of a metropolitan authority controlling the region persisted for the next twenty years until it was realized in 1985.

The idea of a city-region was implemented in many former British colonies. For example, Greater Baghdad in Iraq and Greater Khartoum in Sudan both experienced transformations at the hands of British planners similar to those in Amman.[8] These ideas of city-region and metropolitan authority were not confined, however, to Britain and its former colonies. They were rather universal ideas during the 1950s and 60s. Lloyd Rodwin, a planner from the Massachusetts Institute of Technology (MIT) who worked as a consultant for the Turkish government in the late 1950s and early 60s, presented these ideas as solutions to the problem of metropolitan growth. Starting from a hypothetical notion that we have the ability to anticipate or influence growth, Rodwin suggested the model of concentrated decentralization, which included the creation of several regional centers that compete more effectively with the metropolis. For Rodwin, this model was the appropriate one for developing countries to adopt (Rodwin 1961, 177).

Despite the universality of the city-region idea, its application in Jordan was highly influenced by the British experience. Some American planners did work in Jordan. However, it cannot be said that the United States experience was very influential in Jordan in the 1960s and 70s.[9] The difference between the British and the U.S. experience lies in the power and control vested in the central government in

---

[6]See also Young and Kramer (1978, 12). Young and Kramer argue that the establishment of a metropolitan authority coincided with the objectives of the Conservative government of the time to regain control over the capital of the nation by extending its boundaries to the suburb where large numbers of Conservative voters resided.

[7]Development planning was set up under a 1947 act. This act emphasized *detailed statements of future land-use proposals.* This, according to the planners, did not suit the 1960s. A two-tier system of plan-making could serve the needs in a more efficient way: *first, structure plans containing main policy proposals in broad outline for a wide stretch of territory; second, local plans for smaller areas which would be prepared within the framework of the structure plans as occasion arose, including action-area plans for specific developments* (Hall, 174).

[8]Many of the British planners who worked in Amman in the 1950s worked simultaneously on UN planning missions in Iraq, such as Lock and King, who prepared the first master plan for Amman in 1955. See Berger (1960) and Lock (1958).

[9]Any French influence on planning in Jordan has been dismissed because of their minimal role as a colonial power in Jordan. French approaches to planning were very influential in Syria and Lebanon in addition to most of their former colonies. However, countries other than French former colonies received their share of influence. French planners competed with the U.S. experts in Turkey for a while. In the late 1970s and early 80s, they were more influential in Egypt as well. Khaled Elkhishin (1990) argues that the 1982 master scheme for Cairo treats the city as a "subset of a much larger, heterogeneous urban region." This treatment, he writes, "is analogous to Paris's *poles restructurateurs* concept" (Elkhishin 1990, 126–128).

each case (Rodwin 1970). Edward Banfield (1961) argued that local governments in the U.S. had more tasks to perform than they did in Britain:

> The British ... believe that the government should govern. And [the Americans], although acknowledging that the development of metropolitan areas should be planned, still believe that everyone has a right to "get in on the act" and make his influence felt (Banfield, 98–99).

This difference is what makes the Jordanian experience more British than American. The idea of GA was intended to decrease the power and responsibilities of local governments. The planners' idea of "city-region" meant vesting control in some central authority—in this case, the GAM.

The U.S.'s influence cannot be dismissed entirely, however. Many U.S. experts worked in the country (especially in the 1970s), and U.S.A.I.D. was influential in promoting development in Jordan. The main such effort the work of the Amman Urban Region Planning Group (AURPG), which was formed in 1978 to prepare a comprehensive plan for the Balqa-Amman Region. The AURPG consisted mainly of consultants from Cornell University and a few local planners from the MMRAE and Amman Municipality. A small number of short-term international consultants joined the group at different stages. Arch Dotson, a professor of Urban Planning from Cornell, was the senior advisor for the group. The group envisioned a system of "concentrated decentralization," like that of Turkey. It viewed Amman city as part of a heterogeneous region, within which it has symbiotic relationships with the various parts of the region.[10] This effort was, however, too brief to leave any lasting influence. The AURPG was dismantled, and its work was decreased to an interim report (AURPG 1979).

Like the British planners before them, the planners of the Joint Team in the 1980s accepted the idea of city-region with a two-tier system as a solution for Amman. They had several choices of administrative structure. For example, they could have adopted a one-tier or a multiple-tier system. Indeed, the planners considered these two options. They even singled out several pages in the team's Boundaries Report for each option. In the end, the Joint Team planners decided to adopt a *two-tier* system, despite the fact that they did not think of this option as a "preferred" one.

> Although this may not be a preferred solution, it is compatible with the evolving philosophy of creating a central strategic authority for the whole of Greater Amman, able to control the Master Plan (Greater Amman Municipality 1985, 23).

Thus, the planners' choice was based mainly on their quest for a central authority. This indicates that the idea was so powerful that they did not hesitate to ignore the "preferred solution."

John Calder, the British head of the Joint Team, did not deny the influence of the London experience on him. He even confessed that he was aware of the problems of the idea of city-region in London at the time when the team adopted it for Amman.

---

[10] Arch Dotson stated in an interview with the author on December 14, 1996, that the politics of the context requires a powerful agency to plan. He did not know which agency would take the lead.

But he believed that the differences between Amman and London would allow the idea to succeed in Amman:

> I was skeptical about it … but we were determined, considering the problems of the municipalities …. The various municipalities had a lot of problems: finance problems, no technical [skills], etc. So the idea was to get rid of these municipalities, and replace them by districts … It seemed like a straightforward exercise that needed to be done.[11]

Thus, the idea of GA was a practice in the historical sense. The planners of the Joint Team entered this practice when they chose to implement it. Indeed, the idea itself as a practice and its meaning as a solution for the administrative multiplicity problem were controlling in the sense that they were often repeated and accepted. Even more important is that while the idea was a practice in the historical sense, it was dehistoricized by being removed from the historical context, and it was naturalized by being presented simply as a natural solution to a "preexisting" problem. The two processes of "dehistoricization" and "naturalization" by which the making of GA is characterized hide the quest for control over it. The idea of GA provides certain groups with control while denying it to others. Its realization is a result of this long quest for control. This quest has perhaps always been the motive behind the idea of GA since its inception in the early 1960s. The British planners of the 1960s often complained about the absence of a central authority that could control the peripheries of Amman. So did the decision makers, including several mayors of Amman, on many occasions. A central metropolitan authority would be the device that provides control for those who seek it.[12]

8.4   Greater Amman and the Planners' Quest for "Control"

The idea of a metropolitan authority as suggested by the Joint Team had different meanings to different communities. Each community interpreted it in its own way and reacted toward it accordingly. To many of the municipalities included in the planners' definition of GA, the idea was a threat to their autonomy. A central authority meant eliminating many local governments and restructuring what would be left. It meant reducing the power vested in them by making their decisions dependent on the GAM. Hence, many of the local authorities opposed it. To the planners of the Joint Team, though, the idea had the same meaning it always had: A metropolitan authority meant better control over planning the city-region. They

---

[11] John Calder, project manager. Interview by author. April 8, 1996. London, England.

[12] Over the past fifty years, planners in Jordan have produced many planning acts, or even plans, that reflected a similar line of thinking. There has also been a similarity in the ways by which planning problems were defined. Some differences may have existed in the nature of planning acts, with their constructed problems and their presumed solutions. Still, the mindsets that constructed the problems witnessed stability and produced similar solutions. GA, as the planners of the Join Team envisioned it, is one of those planning acts.

argued that in order for the plan to be efficient, its implementation should be the responsibility of one authority. In other words, the power over the Master Plan could not be shared. And so, the planners insisted that total control over GA should be achieved at an early stage of the plan.

In the end, the Joint Team won, and its idea was implemented. The team's interpretation was in harmony with that of the decision makers. The local authorities' voiced opposition did not find an echo at that level. The Joint Team designed an authority that could control all planning aspects in GA. This control was to be achieved through "procedures of exclusion," to use Michel Foucault's expression (Shapiro 1981, 225). That is, given the fact that the planners were assigned the responsibility of planning by legislation, the deployment of the various objects in planning discourse vests control in the planners, and, given that planning was intended to be a closed field, excludes "non-planners." At the same time, by making certain things objects of planning discourse, planners gain control over those things. And by organizing certain relationships, many parties are excluded from that control (Malkawi 1996). Thus, *objectification* is a practice of control.

The planners of the Joint Team argued that this control was needed. The lack of control, according to Kamal Jalouqa, "has always been the main reason for the failure of the various plans made for the city" over the past years.[13] Accordingly, many of the planning practices included in the GACDP could be viewed as practices of control. For example, by emphasizing the need to prepare the comprehensive plan for the whole GAR instead of Amman city, the planners of the Joint Team made the region an object of the plan. At the same time, by assigning the responsibility of implementing the plan in the GAM, they vested control over the region in themselves. Furthermore, those who resisted the planners' control were viewed as problems. With this in mind, the planners of the Joint Team saw local governments as a problem.

Local governments, like the opposing local authorities, were also against the idea of GA. Indeed, local governments were seen as foes by the planners. They consisted of elected bodies, which the planners believed to be more interested in pleasing their electorates than in *good planning*—a commodity that the planners promulgated the plan with. In other words, the planners believed that local governments were controlled by *special interests*. Waddah Keilani, a member of the Joint Team, emphasized such a belief when he explained how plans got approved:

> The political decision assigns approving plans to a group of councils. The first council is the Local Committee for Zoning which usually includes the members of the elected Municipal Council. Those usually have limited knowledge of planning and they represent their electorates. In other words, they represent the *special interests* they aim to please.[14]

---

[13] Kamal Jalouqa, director of Department of Research and Development, Greater Amman Municipality. Interview by author. April 30, 1994. Amman, Jordan.

[14] Waddah Keilani, Department of Research and Development, Greater Amman Municipality. Interview by author. May 8, 1994. Amman, Jordan.

While local governments were portrayed in the plan as being responsive only to the needs of a small number of special interest groups, the planners themselves, and GAM as a whole, were presented as being more interested in the public welfare. This legitimized the planners' quest for control. Hmud al-Hunayti, a member of the Joint Team, states:

> The municipality of Amman as a service organization is removed from any personal gains and aspires always to provide the best services within the framework provided by the government's plans. The municipality lost money in the process of amalgamation because we inherited the debts of the various municipalities, we inherited unfinished planning, and we inherited social and organizational problems that confused the municipality and changed the whole planning structure.[15]

Accordingly, the GAM planners, the champions of *public good*, should have control over the planning of GA while local authorities should be denied it. This control was achieved through procedures of exclusion.[16] There were two kinds of exclusion in the idea of GA: exclusion from membership and exclusion from power over planning, both aiming at the exclusivity of control. When the planners of the Joint Team redefined GA, they set a new criterion for membership. They decided on the areas to be included in GA. The local governments within those areas were not asked, so to speak, to become members. A decision was made by the central government (responding to the planners' recommendations) and approved by the House of Parliament. On the other hand, membership was denied to many of those who might have wanted it. Membership, however, did not necessarily grant the participants power. On the contrary, some members may have had been more powerful when they acted as autonomous bodies. By becoming members, their power decreased.

The planners of the Joint Team wanted local governments to be powerless. They envisioned a new administrative organization that would have *power over the whole region*. Power, in the planners' vision, could not be shared. It should rather be centralized in an organization, which is nothing but a larger version of the municipality of Amman. Once this question of control was settled through procedures of exclusion, the problem of maintaining this control became the issue. The planners of the Joint Team were speaking from a privileged position. This position was, however, temporal as their power at the time stemmed from their mandate as members of the Joint Team. There had to be a permanent position for the planners to secure their control over planning. Legislative changes were needed to provide a legitimate cover for the plan and assign the control over planning to the GAM. The plan states:

---

[15] Hmud al-Hunayti, director of Tila al-Ali, Khalda, and Umm al-Summaq districts. Interview. June 18, 1995, Tila al-Ali, Amman, Jordan.

[16] Exclusion, often by prohibition, is a practice of power: Those who exclude practice their power and ability to exclude, and those who are excluded lose a legitimate position that they could have gained had not they been excluded. Consequently, they become, relatively, powerless (Foucault 1972).

The new legislation should adopt the Master Plan as the main instrument to be used by the Greater Amman Municipality to guide and control both urban and rural development during the next 20 years. The most fundamental requirement of the legislation is to confirm the Municipality's responsibilities and powers to control the form, content, location and timing of all development within the area of the Municipality (Greater Amman Municipality 1988, 1:14.1).

An authority was needed to grant control and permanence. This authority would enforce, first, the procedures of exclusion and, second, the plan. The authority was supplied by the mayor. His support to the planners made the idea of GA possible after more than 20 years of its first inception, and secured its sustainability for decades to come.

8.5 Conclusion

What turns some things into planning problems and others into solutions? This chapter suggests that a controlling idea, such as the idea of GA, could impose itself as a solution to an imagined problem. Indeed, the whole notion of problem-solution, as it appears in the GACDP, is a form of representation that stresses certain ideas as solutions to which problems were rather constructed. In other words, there are certain acts that are dominating and even they precede the problems they are supposed to solve. Those acts, by being often repeated and accepted, become practices in the historical sense. The idea of GA was one of those acts. For Calder, the head of the Joint Team, it was "a straightforward exercise." He, and his colleagues, were familiar with these situations. There was a perceived problem, and its solution was ready. What made the idea even more dominant was the readiness of some politicians to adopt it. At the top of these politicians was the mayor of Amman, whom Arch Dotson, a Professor of Urban Planning from Cornell University who was a member of the AURPG described as "a very powerful man with connections with the King." Thus, the idea itself was a dominant one that forced itself on almost everybody.

An important issue that still needs to be tackled is the role of the public in this story. This chapter does discuss the local governments' reaction to the idea of a metropolitan government. However, the image of GA as presented in this chapter is that of the Joint Team planners. There are certainly other perspectives that still need to be uncovered. Nevertheless, it is important to note that the public was silent toward the plan when it was being prepared. Despite the high publicity that the plan had, there was little input from the people who would be affected by it at the planning stage. Once it was finished, those became vocal in their criticism. Indeed, one planner argues that perhaps this was what killed the plan, referring to the little impact the plan had on the city due to ignoring many of its policies in selecting development projects for implementation.

It is also important to note that the planners of the Joint Team had little interest in the reaction of the public. They were totalitarian in the sense that they thought they knew the *public good*, worked for it, and believed that they were the ones who could implement it. They believed that they had full understanding of the city and its problems, which should distinguish them from the public, the "people of the street," as one of the planners described the public. Thus, the planners of the Joint Team had a mythic view of themselves. They viewed themselves as symbols of rightness and appropriateness. They presented themselves as the properly *equipped experts* who had the ability to prepare and implement the Master Plan. Indeed, they stressed this image of experts and naturalized it. They presented themselves as national heroes who fight in an "irregular and an unofficially organized body of partisans" (Holston 1989, 68). Thus, they demanded more freedom for themselves as planners in order for their work to be effective. This freedom included, in its many meanings, the freedom to ignore public opinion on the basis that the public was not aware of their needs.

The creation of a city-region with a powerful central authority was, as explained, mainly based on the planners' desire to control planning in the city. This desire was created by the various relationships that the planners had in the society, such as those with the mayor, elected members of local governments, other planners, and, most of all, the king, which was mediated by the various agents of power (namely, the mayor). What made the multiplicity of administrative units a problem, and the idea of GA its solution, is a set of rules. These rules, derived from social and political relationships, allocated authority, responsibility, and control. Indeed, those rules organize the whole planning profession and relate it to the rest of the society. They assign the responsibility of planning to the planners but grant them little control over it. Rather, control is reserved for individuals with "power." In return, the planners aspire for that control. Hence, many planning practices are designed for that purpose only.

What is suggested in this chapter is that the idea of GA is ideological as much it is political. Both power and ideology play a major role in providing choices for the planners and making one choice more plausible than another. They both participate in making a particular idea dominate while other ideas subside. Thus, politicians, perhaps influenced by special interests, may decide on where to open a street or build a mall, where to put a traffic light or build a housing project, or how to define borders and establish authorities. Nevertheless, it is possible that streets are opened, malls are built, and borders are defined only because the planners believed that those were the *right things to do*. In the case of GA, there was a match between what the planners viewed as a good form of governance and what the mayor wanted. Control, in some sense, was vital for both parties. This match between the planners' view and politics is what makes GA an interesting case to study.

The story of GA provokes many questions regarding planning and its limits, but there are also many lessons that can be learned from it. However, the truly essential lesson is the need to raise questions about the social and political implications of certain planning decisions. In doing so, we learn to question the stated meanings of things and the whole notion of *intentions*. This is because, as Foucault puts it, what "counts in the things said by men is not so much what they may have thought, as that which systematizes them from the outset" (1973, xix).

References

AURPG - Amman Urban Region Planning Group (1979) Planned development in Balqa-Amman region, 1981–1985. AURPG, Amman

Banfield E (1961) The political implications of metropolitan growth. In: Rodwin L (ed) The future Metropolis. George Braziller, New York, pp 80–102

Berger M (ed) (1960) The new metropolis in the Arab world. Allied Publishers, New Delhi

Elkhishin KS (1990) Planning for growth in the Cairo region: a strategic management approach modeled on the Paris experience. Dissertation, University of Pennsylvania

Foucault M (1972) The archaeology of knowledge and the discourse on language, translated by AMS Smith. Pantheon Books, New York

Foucault M (1973) The birth of the clinic: an archaeology of medical perception, translated by AMS Smith. Vantage Books, New York

Greater Amman Municipality (1985) Greater Amman comprehensive development plan: boundaries report. Greater Amman Municipality, Amman

Greater Amman Municipality (1988) Greater Amman comprehensive development plan: report 5, final report. Greater Amman Municipality, Amman

Hall P (1975) Urban and regional planning. George Allen and Unwin, London

Holston J (1989) The Modernist City: an anthropological critique of Brasilia. The University of Chicago Press, Chicago

Lock M (1958) Town planning in the Middle East with special reference to Iraq and Jordan, Town and country planning summer school report of proceedings. Town Planning Institute, Bangor, pp 154–176

Malkawi F (1996) Hidden structure: an ethnographic account of the planning of greater Amman. Dissertation, University of Pennsylvania

Malkawi F (2008) The new Arab Metropolis: a new research agenda. In: Shishtani Y (ed) The evolving Arab City. Routledge, New York

Municipality of Amman (1984) Proposal for consulting services for Amman Development Planning, 1985–2005. Municipality of Amman, Amman

Newcombe V (1964) Town and country planning in Jordan. Town Plan Rev 35:238–252

Newcombe V (1967) Planned development in Jordan. Town Ctry Plan 38:412–415

Rodwin L (1961) Metropolitan policy for developing areas. In: Rodwin L (ed) The future metropolis. George Braziller, London, pp 171–189

Rodwin L (1970) Nations and cities: a comparison of strategies for urban growth. Houghton Mifflin Company, Boston

Shapiro M (1981) Language and political understanding: the politics of discursive practices. Yale University Press, New Haven

Turner S (1994) The social theory of practice. MIT Press, Cambridge

Young K, Kramer J (1978) Strategy and conflict in metropolitan housing: suburbia versus the greater London council, 1965–75. Hein, London

Chapter 9
The Endless Challenge of Local Governance in Casablanca

Khadija Darmame, Abdelkader Kaioua, and Eric Ross

9.1 Introduction

To comprehend local governance in the Casablanca metropolitan region, along with its opportunities and challenges, it's imperative to contextualize it within Morocco's overarching governance framework and territorial vision since independence. Indeed, Morocco has placed strong emphasis on issues related to territorial management, and this can be explained by the fact that the nation was going through significant transitions brought on by a variety of factors, including demographic growth, urbanization, socio-spatial inequality, and governance. The driving force behind these territorial-administrative initiatives of the government has been the relentless development of Casablanca (Kaioua 2016).

Casablanca is at the heart of changes in national governance strategy. The city enjoys considerable national economic and demographic clout, and it has the potential to be enticing on a global scale. Thus, it serves as Morocco's economic and political hub. But as a metropolis, it struggles with a number inadequacies and deficiencies that prevent it from operating properly. Over the last decades, planning urban development, where the local government has to provide services to the population, has remained an aspiration for all stakeholders. Due to its complex system of management, the performance of governmental entities is limited. This concern was emphasized by a royal speech in 2013 before parliamentarians, a speech in which King Mohamed VI evoked "a deficit of governance" afflicting the metropolis.

K. Darmame (✉) · E. Ross
Al Akhawayn University in Ifrane, Ifrane, Morocco
e-mail: K.Darmame@aui.ma; E.Ross@aui.ma

A. Kaioua
School of Letters and Social Sciences, Hassan II University, Casablanca, Morocco

9.2   The General Setting for Casablanca' Governance

9.2.1   Key Facts and Figures

Undeniably, the Casablanca-Settat Region (*wilaya* in Arabic) is Morocco's power-house. The region is composed of 7 provinces (substantially rural-urban): Settat, Berrchid, Sidi Bennour, El Jadida, Mediouna, Nouaceur, Benslimane, as well as two prefectures (entirely urban): Mohammedia and Casablanca. The uncontested status of metropolitan area of Greater Casablanca ("Grand Casablanca" in French) is due to its 1615 km^2 area of which 290 km^2 make up the Casablanca prefecture (Casablanca en chiffres 2020). Morocco is easily considered among the leading industrialized countries in Africa (African Development Bank Group 2022). As of 2019, the Casablanca-Settat Region accounted for nearly 32% of the country's National Gross Domestic Production (GDP) and constituted the country's single largest consumer market (Haut Commissariat au Plan 2019).

Renowned as an industrial hub, the region is home to the automobile, electricity, electronics, textile and leather, aerospace, and agri-food industries. Casablanca is recognized as a leading African financial center fully connected to the largest global financial centers (Aljem 2018). Indeed, the financial sector in Casablanca represents 29% of the national banking network and 35% of the insurance-providers network (African Development Bank Group 2022). On the ground, this is manifest in a series of major urban projects: Casablanca Finance City (CFC), the Technopark, and Casa Nearshore Park for the business and technological environment, CasaTramway for public transit, and the Port of Casablanca.[1] The population of the municipality of Casablanca was estimated at 3.84 million in 2022, with about 6 million in the Greater Casablanca metropolitan area (World Population Review 2022). According to official predictions, the population of Greater Casablanca was expected reach 5.5 million inhabitants in 2030 (Kaioua 2016), but this seems already to have been surpassed. Regardless, it is classified as the most populous city in the Maghreb region (Morocco, Tunisia and Algeria). It is the eighth largest city in both the Arab world (Edelman 2023) and in Africa.

Casablanca is unique for its exceptional peri-urbanization dynamics, and this has produced rapid population growth in peri-urban areas to absorb the surplus (Chouiki 2010). Casablanca's sustained demographic growth generates a continuous process of urbanization. Lissasfa and Sidi Maarouf are exemplars of peri-urban municipalities fully integrated into the city's contiguous urbanized area (Fig. 9.1).

Chouiki (2017) distinguishes three distinct rings of peripheral urbanization surrounding Casablanca's core:

---

[1]The largest port in Morocco, accounting for 23% of national traffic according to Ministère de l'équipement et de l'eau. Available at: http://www.equipement.gov.ma/Carte-Region/RegionCasablanca/Presentation-de-la-region/Monographie/Pages/Monographie-de-la-region.aspx

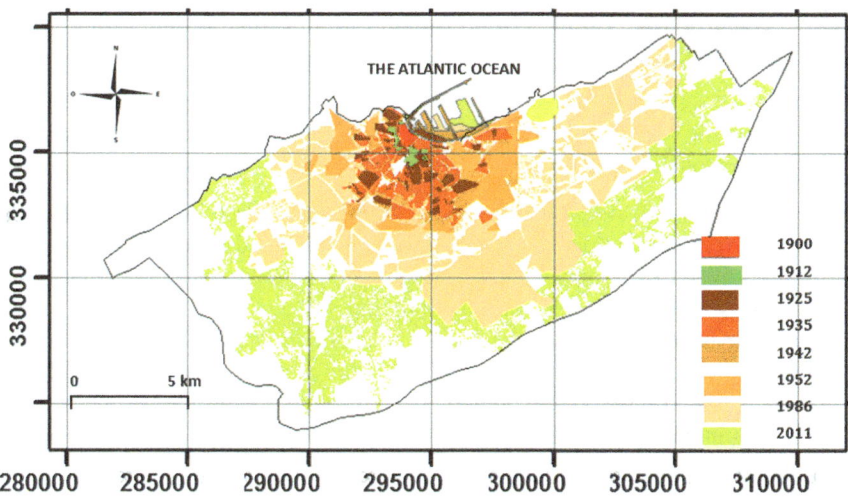

Fig. 9.1 Map of urban expansion of Casablanca. (Source: Loukili et al. 2015: 262. Published with permission)

1. The first ring, immediately adjacent to the core for 1 to 5 km is a discontinuous residential area of unregulated residential clusters around well disguised warehouses.
2. The second ring, between 10 to 30 km from the city center, is a classic rural-urban mix that hosts a range of agricultural and industrial activities. This area is marked by a lack of regulation and is not properly developed.
3. The third ring extends to the outermost edges of the Greater Casablanca area. It retains characteristics of both rings although it is mostly rural, inhabited by households that are unable to afford more urban rents.

Beyond spatial expansion, it is important to consider the social dimensions of Casablanca's growth, particularly the social and spatial inequalities that have accompanied its urban expansion and integration (Beier 2020). Indeed, social divisions are clearly visible within the metropolis' suburban fabric. If the State's discourse tends to be promote integration, as stated by Nejmi (2018), urbanization practices generates social exclusion of low-income groups. The city has progressively integrated higher income residents, driving the living cost higher, while pushing the poor towards the periphery, where socio-spatial differentiations continue to worsen. This results in a waste of resources, including financial resources and human capital, accompanied by chaotic city management. Municipalities within the metropolitan area are confronted with the greatest challenges. There remains an urgent need for major investment in infrastructure, transportation, equipment, and social housing (Nejmi 2018; Kaioua 2016).

9.3   Casablanca' Local Governance: A Perpetual Readjustment for Territorial Efficiency

According to the Constitution of 2011, local governments (LGs) are local authorities with legal personality and financial autonomy, which are divided into three levels:

1. **Regions**: Morocco is composed of 12 regions that attained the status of LGs in 1996. A region is headed by a *wali* (viceroy) appointed by the king. The wali presides over a Regional Council consisting of one representative appointed by each of its constituent prefectures and provinces.
2. **Prefectures and Provinces**: the regions are divided into Prefectures, which are predominantly or exclusively urban, such as the prefectures of Rabat and Casablanca, while Provinces are mainly rural. Both entities are headed by a governor, also appointed by the king. The governor presides over a Provincial Council consisting of one representative appointed by each of its constituent communes.
3. **Communes**: the commune (municipality) is the lowest level of LG. The communes are urban or rural and the law sets a distinction between the two categories. A commune is headed by a mayor (or president of the municipal council), elected to office by the councillors, themselves elected by the citizens.

Walis and Governors are appointed by the king to represent the central administration in the regions, provinces and prefectures. Their competences are rarely defined with any precision but as representatives of the king and the Ministry of the Interior, they concentrate more powers than the elected authorities at the prefectural and provincial level (Bergh 2013). There is no hierarchical relationship between the various territorial levels and no local authority exercises any power over another. According to the Constitution, the relations between these entities are organic, based on coordination and cooperation. However, the municipal council, being directly elected by the citizens, represents the basic constituency of provincial, prefectural and regional councils.

In Casablanca, the question of building an efficient system of local governance is a challenge that has been consistently pursued for the past half century. Public authorities are constantly seeking to improve intervention tools in their territory to meet the needs of its economic growth and its socio-spatial dynamics. In fact, the metropolis offers an example full of lessons in this regard. It constitutes a sort of experimental laboratory for other Moroccan cities as the changes introduced in Casablanca's metropolitan management system have subsequently been generalized to all large cities across the country.

For almost half a century, Casablanca has been searching for an effective management system. Following independence, Morocco adopted its first municipal charter in 1960, organizing the management of the city of Casablanca within the framework of a single prefecture and a single municipality. During the first two decades (1956–1976), power in the city was shared between an elected assembly with a consultative role and the local authority, represented by the appointed governor who held real power and was responsible to the central administration.

9.3.1 The Early Beginnings of a Governance Strategy, the 1970s and 80s

The governance process specific to Casablanca began in the mid-70s when the government decided to provide the city with a special administrative regime to "increase efficiency in the management of its affairs." The crisis that was impacting the city made this action extremely urgent (Kaioua 1988). The adoption of the second community charter on September 30, 1976, framed the administrative and financial organization of local authorities and introduced a new distribution of jurisdiction between elected representatives and the appointed authorities. The municipality of Casablanca was subdivided into five urban municipalities: Ain Diab, Ain Sbaa, Ain Chock, Mers Sultan, and Ben M'Sik.

Starting in the 1980s, the substance of communal power was wielded by officials appointed by the territorial administration. Yet, despite the centralization of power and great spatial fragmentation, local authorities were unable to meet the expectations of a rapidly growing population. This was a real turning point. Elected council members began to directly manage their municipalities, and the powers of both the municipal councils and their presidents expanded. The five municipalities were headed by a newly created entity set up to coordinate and manage common services that required inter-communal solutions. This was the Urban Community of Casablanca (CUC). This organization engendered great hope when first established, and it did enable substantial progress to be made in framing and managing the problems of the city.

The respective powers of the territorial entities, the urban municipalities and the larger urban community, were not sufficiently delineated. This generated numerous difficulties and the experiment lasted only five years. The charged social movements (urban riots) that shook Casablanca in June 1981 marked a new phase in the city's political organization. They accelerated awareness of the extent of its problems, and public authorities committed themselves to taking control of the municipal affairs. This inaugurated the third stage of intervention, driven by a tense social climate that resulted in a deep reform and adoption of new management tools. During the 80s, administrative and technical management structures were redesigned. These were then periodically readjusted over the succeeding 20 years.

Starting June 1981, a dual process affected both territorial administration and municipal management. Though still one single prefecture, Casablanca consisted of five distinct municipalities, to which a six was added when the city of Mohammadia was included. This marked the birth of "Grand Casablanca" (Greater Casablanca). That year, a higher coordinating structure was created, the Wilaya du Grand Casablanca. This territorial reorganization was accompanied in 1983 by a complete redesign of the administrative map, which was subdivided into 23 small entities called "districts." A new urban master plan, adopted in 1984, aimed to control urbanization and solve problems that had been accumulating for years. The marked increase in number of decision-making bodies and the rise in the number of players involved in territorial management led to the creation of a technical unit to control the urban development of Greater Casablanca. As an outcome, the first "agence urbaine" (urban planning agency) was created in October 1984. It was placed directly under the Ministry of the Interior and headed by a Governor Director General.

This squeezing of the administrative distribution was accompanied in the early 1980s by the decision to equip Casablanca with its third urban development plan since the beginning of the twentieth century (Kaioua 2016). This major revision reflected, first and foremost, concerns over territorial security. The experience lasted for almost 20 years, from 1981 to 2002, and had several positive aspects, such as bringing the administration closer to citizens, and the emergence of a new elite of elected representatives and public affairs managers. At that time, Greater Casablanca had a total of 1100 elected representatives. Did this large number of advisers make the communes more effective? The assessment of the experience revealed results below target. The undeclared aim of increasing the number of communes was to limit their political influence and management efficiency, so that "no politically organized force could have any kind of preponderance or dominance at the level of the agglomeration, the city's urban area, as a whole" (Naciri 2000).

Since the adoption of decentralization following independence, Morocco has been regularly updated its territorial administrative divisions to accommodate the country's changing social and politico-economic scene. Although territorial entities were created in the 1960s, they were not adopted effective on the ground until the status of local collectivity was implemented in a 1992 constitutional amendment. Morocco adopted decentralization in its laws and policies starting in 1959, but the actual application only got underway in 2000s. The aim was to develop the country across sectors with the principle of a democratic equal progress. The changes in administrative subdivisions resulted in a reorganization of its regionalization model, the latest being in 2015, which in turn affected the number of provinces and prefectures. How this impacted Greater Casablanca can be visualized in Fig. 9.2.

9.3.2   2000s the Attempt to Increase Casablanca' Governance Efficiency and Effectiveness

At the beginning of the 2000s, after many successive revisions, the Casablanca metropolitan area covered nine prefectures and 35 communes. The new quest was for efficiency in the management of Casablanca's affairs. A royal speech delivered in Casablanca at the beginning of the reign, on October 12, 1999, set the agenda for a "new concept of authority." "Our territorial administration must henceforth specifically focus on areas of particular importance and priority, such as environmental protection and social action, and use all means to integrate the underprivileged into society and ensure their dignity." This speech committed the government to implement reforms likely to strengthen the decentralization process in the country. At the end of 2002, a new communal charter was adopted. It acted upon a request that had

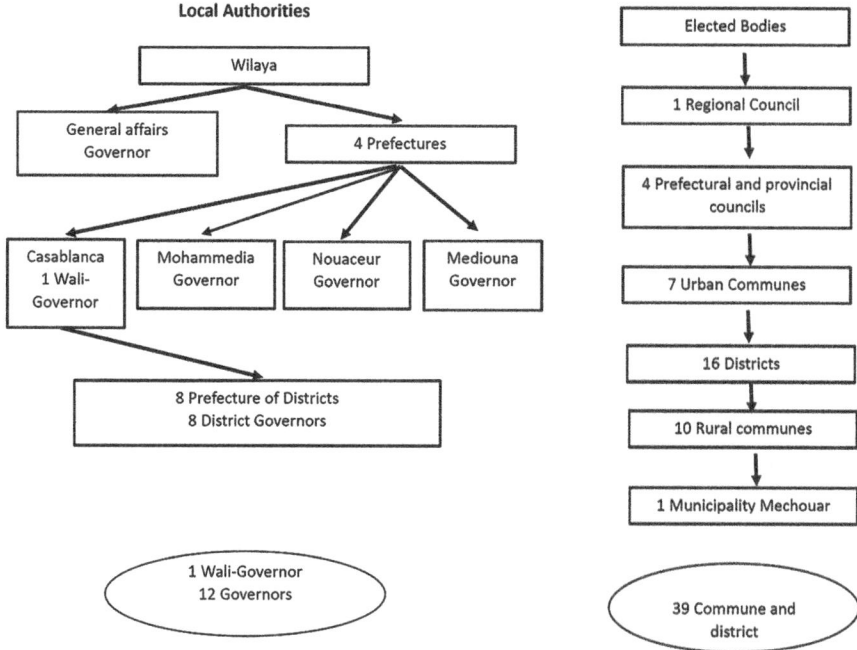

Fig. 9.2 Territorial and Communal Administration of Greater Casablanca. (Source: Kaioua 2016)

long been expressed by Casablanca's stakeholders, that the fragmented administrative geography that governed major cities be replaced by a single city council.

Starting September 2003, the territorial management of Casablanca was unified once again. The "Urban Community" was transformed into an "Urban Commune," or municipality, reestablishing the unity of the city. The administrative map of Greater Casablanca was remodeled. The nine former prefectures were replaced by two (Casablanca and Mohammadia) and two new provinces were created (Nouaceur and Médiouna). However, this reorganization introduced a new administrative level exclusively applied to Casablanca. These are the "district prefectures." A soon as Casablanca prefecture was reinstated, it was subdivided into eight district prefectures, each by governors under the authority of the Wali of the region.

In terms of how the municipality actually functions, there is now a single "commune" for the entire city, but it comprises 16 districts, each represented in the municipal council by its president. Greater Casablanca is now governed by 12 governors and one Wali, and it covers 39 local authorities. This reorganization has permitted enhancement of both technical planning and implementation structures. Thus, on the eve of launching Morocco's "advanced regionalization," 115 distinct entities still made up the governance institutions of Greater Casablanca. In view of the many challenges to be addressed, the complexity of the institutional governance does not make effective public action in Casablanca easy. This is the cause of major dysfunctions.

9.4   Assessing Casablanca' Governance Initiatives

9.4.1   Major Challenges vs Minimum Opportunities

The Communal Charter of 2008 devolves political powers to municipalities, thereby reducing the supervisory powers of the Ministry of the Interior and its officials over municipal affairs (Clark 2015). Thus, the municipalities have the possibility to set matters of local concern in the following areas:

- Economic and social development plans that should be in line with the guidelines and objectives set at national level, (Article 36 of Law 17.08),
- Financial affairs, local taxes and management of collective land,
- Urban development and territorial management,
- Creation and management of local public utilities,
- Health protection, hygiene and the environment,
- Equipment and social and cultural affairs,
- Cooperation and partnership with private sector, and
- Financial contributions to the semi-public companies operating at the local level, voting its budget, reviewing and approving their financial statements.

9.4.2   Excess of Monitoring and Interference of Institutional Actors

Although many reforms took place to enhance the autonomy of LGs and in particular the municipalities, the local governance system is still under the tight control and supervision of The Ministry of the Interior and the Ministry of Finance. They have to endorse and ratify the most important decisions and deliberations made by the municipalities on local matters. The Ministry of the Interior acts via its General Directorate of Local Collectivities, which maintains a double hold on municipalities, in technical and juridical matters, and over budgetary matters as well (Nebie and Tbitbi 2021)

After ten years of daily operations under the new structure (2003–2013), a series of problems emerged in their functioning which hindered their effectiveness. At the level of municipal management, the single city system failed to frame the behaviors of players. The conflicts of authority reflected the relations between the various councils in the management of strategic sectors such as urban matters, hygiene, or infrastructure and equipment. The lack of precise prerogatives distinguishing the municipal council and the various district councils explain the contradictions and overlaps of authority and are the origin of the blockage in the functioning of the city.

9.4.3 Overlapping and Competing Jurisdictions of Responsibilities Between LGs and the Central Government

The municipalities are in charge of public utilities through autonomous municipal administrative corporations (la Régie Autonome, in French). The corporations manage the water and wastewater sector, electric power, lighting of urban public space, transport, trash collection, and coordinate transversal infrastructure projects. Other important public services such as health care, hospital services and education, are still in the jurisdiction of the central government. The municipalities play the role of regulator of Public-Private Partnerships (PPPs) to manage urban utilities. The PPPs were required by the top of the hierarchy (the Royal Palace and Ministry of the Interior). Theoretically, PPPs should be accountable to the municipalities, who must monitor their performance indicators. Thus, the private companies involved in delivering public utilities in Morocco have markedly interfered with the municipality's responsibilities and stripped them of some of their crucial tasks.[2] Furthermore, the municipal council members lack the skills and tools required to regulate and monitor such contracts.

Various ministries, such as the Ministry of Transport, Ministry of Urbanism and National Land Settlement, Ministry of Housing and Urban Policy, are represented at the local level by their executive directorates. They play a significant role in management of the city and compete with the municipal administrations. These directorates ensure that national plans and policies are adapted at the local level and monitor their implementation. These Directorates are also responsible for the execution of projects that the municipalities cannot due to the lack of sufficient financial resources. We can cite the example of the Ministry of Urbanism and National Land Settlement, via its Directorate of Regional Planning, that ensures the coherence of public policy by monitoring and evaluating its implementation, providing information and analysis to local decision-makers and supporting development initiatives (Table 9.1).

The multiplicity of institutional actors involved in managing the city's services are supposed to coordinate and cooperate with each other. However, for some matters, there is an overlap between the central government and LGs. This is particularly acute in social services (health, education, housing), where jurisdictions overlap. Both levels of government develop the same socio-economic projects and design the same social facilities leading to a duplication of efforts. The amendment of the Communal Charter of 2009 tried to alleviate this issue. Henceforth, as part of its advisory role, the municipality is to be informed of any project to be carried out in the territory of the municipality by an agency of the central government or any public

---

[2] Some of the main international companies involved in the water sector include: Suez Environment in Casablanca, and Vivendi in Rabat, Tangier, and Tetouan.

Table 9.1 Service delivery responsibilities by type of entity

Entity Service	Municipalities	Prefectures & Wilaya	Delagted ministires	Public/Private Partners
Basic Urban Infrastructure Services				
Roads	√		√	√
Sewage	√			√
Green spaces, parks	√			
Cemeteries	√			
Water network	√			√
Basic Urban Services				
Garbage collection	√			√
Fire fighting		√		
Traffic	√		√	
Public transport	√		√	
Street cleaning	√			√
Environmental protection	√	√		
City Development Services				
Master plan	√	√	√	
Implemntation of city development	√	√	√	
Plan city development	√	√	√	
License & instruction control	√	√	√	
Regulatory Services				
Health regulations for food & beverages	√	√		
Wholesale & retail	√	√		
Slaughterhouses	√	√		
Permits for businesses and hotels			√	
Social and Cultural Services				
Entertainment	√	√		
Preservation of historical sites, buildingings & maintenance, sports facilities	√	√	√	
Libraries & cultural centers	√		√	
Health			√	
Education			√	
Social affairs (including care for elderly and children)			√	

Source: Data collected in the municipality of Casablanca, 2023

body. LGs can request the cancellation of the central authority's decisions in the case of abuse of power (Law no. 41-90 establishing administrative tribunals). Overlapping jurisdiction and interference in municipal affairs are mainly caused by clauses in the legislative texts concerning the functioning of LGs and ministerial directorates. Examples include "the LGs have the right to intervene in all the commune's affairs and matters," and "prefecture or region affairs" are the jurisdiction of the regional council or prefectural council." References to responsibilities are vague and ill-defined for each entity.

In order to cope with this situation, the General Directorate of Local Collectivities introduced Local Development Companies (LDC).[3] These are limited-liability public companies currently managed by the organic Law on municipalities known as Law no. 113-114 of July 7, 2015, which revoked the municipal charter of 2002.[4] They were established to manage services, equipment and activities that fall under the authority of municipalities and LGs and operate within the municipal council's territorial jurisdiction. They are useful tools to address deficiencies in public services provisioning as they are independent entities with flexible management and procedures. They set the objectives, then develop, implement, manage, and monitor the delivery of local public service projects delegated to them by the municipality. The first LDCs to acquire contracts were in some municipalities of Casablanca, Rabat, Temara, and Marrakech. These related to water, sanitation, urban transport, cultural and social activities, urban planning, and parking management.

Casablanca has seven LDCs active in various sectors[5]: Casablanca Prestation (delivery of public services), Casablanca Aménagment (urban planning), Casablanca Baïa (waste management and sanitation), Casablanca Events & Animation (management of cultural events), Casablanca Transport, and Casablanca Patrimoine (urban heritage preservation). While this model has been presented as an efficient way to improve services and address financing and human resource issues, some local elected representatives claim that delegating urban service management to LDCs is an attempt to weaken local democracy and accountability, with the goal of reducing the role and scope of elected councils in fulfilling of their duties.[6]

This complicated situation has not allowed the building of a shared vision for the future of the metropolis, or the emergence of skilled managers capable of coping with development challenges. The lack of collaboration between territorial actors has led to a deficit in terms of convergence and implementation of decisions. The management ineffectiveness and the weakness of financial resources, associated with dispersal of decision-making centers, have only contributed to the increase of blockages in managing matters within the Greater Casablanca. Social inequalities and forms of exclusion, in terms of social services and unemployment, are increasing. This exclusion reflects the failure to implement inclusive development strategies, preventing any attempts to improve the governance system to foster the

---

[3] Called Sociétés de Développement Local in French.

[4] For further details, see the portal of Local Governments https://www.collectivites-territoriales.gov.ma/fr/societes-de-developpement,

[5] Casablanca Municipality's website portal: https://www.casablancacity.ma/fr/article/243/societes-de-developpement-local

[6] Interviews carried out in July 2023 at the municipal council of Casablanca.

emergence of a collective synergy project. The King paid particular attention to this situation in a speech to the parliament on October 11, 2013.

> (....) But one might wonder why, unlike many other cities, Casablanca, one of Morocco's wealthiest cities, is not enjoying the real progress that its citizens, both men and women, have been hoping for. Should it, however, continue to be a city of such huge contrasts, to the point where it has become one of the worst examples of local mismanagement? (....) In a nutshell, the problems of the economic capital of the country stem mainly from governance.[7]

It seems that the government trusts the private sector, more than its own officials and experts, to lead projects. The case of Casablanca's Marina project is an interesting example. This strategically located show-piece real-estate development project was entrusted to private companies under the supervision of the local authorities. Faced with pressure from local governments, including Casablanca's Urban Agency (Barthel 2010: 72–75). We should question the causes of the governance stalemate. Is it due to the relationship of LGs to the private sector entities responsible for the implementation of projects? Or does it result from the constant reconfiguration of the administrative map? We can trace the blockages back to the changes in the rules & policies and dynamics of governance which started with the adoption of decentralization. What impact did centralization have on the policy agenda? According to Lamia Zaki, a Moroccan urban governance research,

> (T)he authoritarian management of Local Governments (is) based on the alliance built after independence between the monarchy and rural elites to counter the influence of urban and partisan elites (....). This approach led to the creation of domesticated local elites (...). Researchers have pointed to the centralized and often brutal management of the Moroccan territory, and of cities in particular, in a context of rapid urban growth (...). A second analytical perspective highlights how decentralization has been presented and used since the 1990s by the Moroccan government as a tool to implement "democratization reforms." The relative opening up of the political field in the late 1990s and early 2000s led to the emergence of new elites in different fields (entrepreneurship, real-estate, political parties, civil society, etc.), who have used their local base to claim rights and/or a political role at the local or national level. The third analytical perspective links decentralization to ("good") governance reforms that focus more on the rationalization of resources, effective investment and the respect of management rules to promote local development rather than on representative democracy." (Zaki 2019)

Following the lack of performance in implementing many projects in Casablanca, a royal strategy for urban development was launched to push for progress. It can be translated as a framework allowing the city to have the resources and reasons to overcome the obstacles and establish an inclusive resilient development model, but to also build a proper environment for investors.

---

[7] King Mohamed VI's speech at the opening of first session of third legislative year of the ninth legislature, October 12th, 2013.

9.5 2015–2020 Action Plan in Response to the Governance Deficit in Casablanca

In 2014, a five-year strategic development plan (2015–2020) was elaborated with a budget of 33.6 billion DH (about 3 billion euros), more than half of it is devoted to the urban transportation and mobility sector (Safir 2015). Four levers or enablers were initiated for the implementation of the plan:

1. Effective governance for a better distribution of management responsibilities among the various stakeholders.
2. Funding, to act and adapt institutional measures or arrangements to implement major projects.
3. Effective management by focusing on the revision of administrative procedures to enhance relations between businesses and citizens.
4. Territorial Marketing to reinforce the international positioning of Casablanca.

These are aimed at easing the territorial expansion of Casablanca, formerly a cramped metropolis.

The new dynamic triggered by advanced regionalization offers opportunities for coherence and convergence of public policies at all levels. It is about leaning on the realities and potentials of territory to enable development options. The recent enlargement of the regional territory and the transfer of wide range of prerogatives to local actors created the conditions to remove the obstacles hindering Casablanca's spatial growth. This is a major project that can only be carried out by public authorities because of the remarkably close ties between Casablanca and the rest of the country. Thus, it is possible today to promote a new development model for Casablanca, one that enhances the international stature of the metropolis while fostering inclusive development for its populations. Nonetheless, this still requires major change in intervention modes at various levels so as to align development programs to the aspirations of different social strata.

The territorial administrative reforms have allowed the upgrading of Casablanca to the rank of metropolis (Tomas 2020: 9). The four levers are categorized based on the degree of institutionalization in terms of decision-making and policy implementation. According to the current policy approach based on the 2011 constitution, Casablanca has a metropolitan government where local authorities have significant responsibilities for managing the city and its surrounding provinces. There are a number of city-regions that Casablanca shares similarities with and can learn from, such as French metropolises, Istanbul, Lisbon, and Manila. The French model of metropolitan governance and spatial management struggles with the power relationship between its institutions (Girardin 2022; Bury 2003). However, France transitioned from an "organizational hierarchy that privileged the central city," to having "several territorial authorities actively engaged in shaping their metropolitan future" (Nicholls 2005: 797). This transition empowered new stakeholders to engage with the system, demanding more power in order to do so. However, cooperation is not enough and there is also a need for coordination to achieve equity and balanced metropolitan governance.

In Southeast Asia, metropolitan regions suffer from lack of trust in governmental institutions and question their ability to deliver megaprojects, which explains why they opt for private companies to carry out such operations (Shaw and Satish 2007). Consequently, there is a gap between the plans being implemented and the types of projects needed, projects the private sector cannot deliver as they fall within the power of the government. Spatial inequity requires a bottom-up approach to governance as it builds on the needs of the people, and the feasibility of the projects in terms of infrastructure, budget, and governance (Stephens and Wikstrom 2000) . Another common problem with the territorial administration of metropolitan areas is the adoption of the latest governance models, which tend to trend globally. These rarely take into consideration the characteristics of individual cities, as is the case in Istanbul. There is relentless pressure to integrate a chimeric "global city" socio-economic project which leads to the "cut and paste" of development strategies (Enlil 2011: 6).

In Portugal, a metropolitan governance model based on the principle of coordination has been adopted. Like Casablanca, historically, Lisbon has suffered from a lack of coordination between a multitude of territorial subdivisions of various scales. Each one would plan for growth without regard for its neighbours, leading to great incoherence. This governance system relied on "a system of voluntary inter-organizational co-operation involving various bodies having different geographical extent, having severe problems of political and technical co-ordination, and involving complex urban decision-making processes, all of which constitutes a rather chaotic administrative geography, full of duplication and confusion" (Nunes Silva 2002: 29). This model centralized the governance of Lisbon, all the while weakening the metropolis in the relation to economic stakeholders.

9.6   Casablanca Between Laws, Plans, and the Royal Initiatives

Over the past two decades, a succession of development strategies has been initiated at various levels in the Casablanca region. Assessment reports have revealed results well below declared expectations. Strategies cannot achieve the desired results if they are not integrated around coherent public action. The very first condition of "good" territorial governance is that all stakeholders be accountable. Improving implementation tools is another condition of an administrative system's effectiveness. The lack of land regulation policies, laissez-faire attitudes towards land speculation, disregard for approved plans, and lack of consideration for community needs and desires, all these factors will impede the ability of public authorities to effectively control the growth of a metropolis—no matter how good the official plans are.

Another important factor for effective public administration remains almost absent in Casablanca's case, namely accurate knowledge of the territory and the monitoring of its evolution. Casablanca should acquire independent, highly skilled scientific observation and analysis instruments to monitor changes taking place in real time, and to anticipate solutions.

9.6.1 Casablanca's Governance Stalemate: A Continuous Struggle

In Morocco, as in many other countries, the municipalities and local governments have been assigned a greater role in pursuing their own development strategies without the institutional reform nor capacity building required for the success of these strategies. This leads to overwhelming the municipalities with responsibilities with neither accountability plans nor evaluation mechanisms in place. Although the scope of their responsibilities has increased, the top-down involvement of the Ministry of the Interior is dis-empowering (Zaki 2019). The 2011 constitution was supposed to open the path of legal reforms facilitating decentralization of decision-making. However, we can argue that effective decentralization requires that local authorities have "significant administrative freedoms, and decentralization reforms need to be more closely linked to more efficient fiscal arrangements for municipalities as they develop their own resources." (Zaki 2019) Empowerment local governments should resolve the issue of top-down governance practices. This, in turn, will stop the process of continuous tampering with the place-identity of citizens.

Official discourse, from the king on down, gives voice to the aspiration for local, democratic, inclusive, equitable development. In a 2006 speech delivered to the National Forum of Local Collectivities in Agadir, King Mohamed VI praised the work and emphasized the role of municipalities in the development of local democracy. He emphasized the need for better governance models built on cooperation between the collectivities and the central government over national issues like rural-urban migration, poverty alleviation, and unemployment.

In the case of Morocco, and particularly Casablanca, the king has always been the primary "change agent." His programs and visions are adopted in all action (Hachimi Alaoui 2017: 192). Royal involvement overrides any existing management model and prejudices the selection of the stakeholders who would deliver the projects. Local officials find themselves "excluded from the process in advance," while "a time for state interventionism" gapes, allowing national agencies to take the upper hand. (Hachimi Alaoui 2017, p. 2–5).

The fact that visions emanating from the summit of political power undergird every major project raises questions about how this top-down approach to development squares with the decentralization plans articulated for the country. Obviously, it does not. It is instructive to look at the position of the World Bank vis-à-vis

Morocco's top-down governance approach when it assessed a request from Casablanca's City Council for a loan (to be guaranteed by the Moroccan government). In its appraisal document, Casablanca is described as a success story of collaboration between local, regional, and central governments to fulfill the royal vision. (World Bank 2017) The city is portrayed as a model to follow when it comes to policy making. Its proposed reform plan corresponds to Sustainable Development Goals and solves governance issues through central actions and strategies.

9.7   Conclusion

The last half century has been marked by a neoliberal vision of the city. This vision now permeates every aspect of policy design and development planning across the Arab world. Morocco is no exception. King Mohamed VI has used urban development to pave the way for his national developmental agenda. Governmental discourse reduces urban planning and development to showy megaprojects, tools that will "boost" the metropolis, making it investor-friendly in a highly competitive global market. The implementation of the neoliberal vision through mega infrastructure projects is integral to the royal development strategy. It is responsible for the establishment of Regional Investment Centers, for the creation of Tangier-Med Port, of the Bouregreg Valley Development project in Rabat, Casablanca Finance City, Casa-Marina and other such projects in major cities. The driver of all these projects has been the top-down ability of the regional walis to impose royal initiatives on municipalities in their jurisdiction in the name of results-oriented efficiency, disregard any plans the latter may have developed through bottom-up participatory processes. Royal visions are implemented by the Ministry of the Interior through viceroys and governors. According to the constitution and official decentralization policies, the roles of these royal appointees should be limited to oversight of the work of local governments. Yet, their prerogatives as agents of royal will create space for all manner of interference which sidelines the work of municipal administrations without necessarily guaranteeing efficiency. There were budgeting issues, administrative blockages and problems of timely end-product delivery in each of the megaprojects listed above.

In this chapter we have laid out how Morocco has developed a decentralized system based on democratic institutions to support its governance model. Decentralization policies were adopted to remedy the governance flaws of the post-colonial state-building era. Theoretically, decentralization must be accompanied by democratization at the scale of localities. Oftentimes though, in practice, the two processes are at odds. As power has de-concentrated, effective governance has fallen through the cracks, empowering private sector investors rather than local citizens.

References

African Development Bank Group (2022) Africa industrialization index 2022, Making a Difference. Available at: https://www.afdb.org/en/documents/africa-industrialization-index-2022

Aljem S (2018) Politique des grands projets et gouvernance urbaine à Casablanca. Les Cahiers d'EMAM 30. https://doi.org/10.4000/emam.1446

Barthel P-A (2010) Casablanca-Marina: Un nouvel urbanisme marocain des grands projets. Autrepart 55(3):71–88. https://doi.org/10.3917/autr.055.0071

Beier R (2020) The world-class city comes by tramway: reframing Casablanca's urban peripheries through public transport. Urban Stud 57(9):1827–1844. https://doi.org/10.1177/0042098019853475

Bergh S (2013) Decentralization and Local Governance. In The EuroMed Region Conference, http://cor.europa.eu/en/activities/arlem/activities/meetings/Documents/associations-bergh-local%20governance-euroded-en.pdf

Bury J-C (2003) Métropoles et Structuration du Territoire (Note 137; p. 4). Section des Économies Régionales et de l'Aménagement du Territoire. Les éditions des Journaux Officiels, Paris, avril 2003, available at: https://www.vie-publique.fr/rapport/28041-metropoles-et-structuration-du-territoire

Casablanca en chiffres (2020) CasablancaCity.ma. Available at: https://www.casablancacity.ma/fr/article/230/casablanca-en-chiffres

Chouiki M (2010) La Periurbanisation Metropolitaine au Maroc : Diversité des formes et similitude des fondements. In: Dlala H (ed) Les marges périurbaines en Tunisie et en Méditerranée. Publications de la FSHS, Tunis

Chouiki M (2017) Un siècle d'urbanisme. Le Devenir de la Ville Marocaine. L'Harmattan, Paris, Cahiers de Géographie du Québec, Érudit, available at: https://www.erudit.org/en/journals/cgq/2018-v62-n175-cgq04385/1057090ar/

Clark J (2015) The 2009 communal charter and local service delivery in Morocco (2015). Program on governance and local development working paper #2, available at https://ssrn.com/abstract=3715521 or https://doi.org/10.2139/ssrn.3715521

Edelman D (2023) Managing the urban environment of Casablanca, Morocco. Curr Urban Stud 11(96). https://www.scirp.org/journal/paperinformation.aspx?paperid=123656

Enlil ZM (2011) The neoliberal agenda and the changing urban form of Istanbul. Int Plan Stud 16(1):5–25. https://doi.org/10.1080/13563475.2011.552475

Girardin A (2022) De la désindustrialisation à la vitrine métropolitaine: un quartier du Havre à l'heure néolibérale. Géoconfluences. https://geoconfluences.ens-lyon.fr/informations-scientifiques/dossiers-regionaux/la-france-desterritoires-en-mutation/articles-scientifiques/desindustrialisation-metropolisation-le-havre

Hachimi Alaoui N (2017) A 'time' to act: the 2015–20 development plan for greater Casablanca. Revue Internationale de Politique de Développement 8:189–219. https://doi.org/10.4000/poldev.2321

Haut Commissariat au *Plan* (2019) Note d'information relative aux comptes régionaux de l'année 2019. Available at: https://www.hcp.ma/Les-comptes-regionaux-de-l-annee-2019_a2735.html

Kaioua A (ed) (1988) Découpage administratif du Grand Casablanca, Atlas de la Wilaya du Grand Casablanca. Publication GREC/Université Hassan II Casablanca/URBAMA Université de Tours

Kaioua A (2016) La métropole casablancaise à la recherche d'une gouvernance efficiente et durable. In: Gouvernance durable des villes du Maroc. CESE (Conseil Économique, Social et environnemental), Rabat

Loukili I, Ghafiri A, El Moutaki S, El Hakdaoui M (2015) Monitoring the Spatio-temporal urban expansion of Casablanca City (Morocco) on more than a century using GIS and remote sensing. Eur J Sci Res 135(3):258–267

Naciri M (2000) L'aménagement des villes et ses enjeux. In: Attrait de la ville, crise de la cité. Publication Club Convergence 21, Rabat

Nebie E, Tbitbi E (2021) The challenge of performance in the governance of local authorities in Morocco: Issues and perspectives. International Journal of Accounting, Finance, Auditing, Management and Economics, 2(4):286–301. https://doi.org/10.5281/zenodo.5134679

Nejmi E H (2018) Casablanca: De la ville projet urbain à la ville projet social: Quelle lecture des documents d'urbanisme. Revue Espace Géographique et Société Marocaine 22 available at: https://revues.imist.ma/index.php/EGSM/article/view/12547

Nicholls WJ (2005) Power and governance: metropolitan governance in France. Urban Stud 42(4):783–800. https://doi.org/10.1080/00420980500060426

Nunes Silva C (2002) Governing metropolitan Lisbon: a tale of fragmented urban governance. GeoJournal 58(1):23–32. https://doi.org/10.1023/B:GEJO.0000006567.09464.56

Safir K (2015) Plan de développement stratégique de Casablanca 2015–2020: Quel Mode De Gouvernance Pour La Métropole?

Shaw A, Satish MK (2007) Metropolitan restructuring in post-liberalized India: separating the global and the local. Cities 24(2):148–163. https://doi.org/10.1016/j.cities.2006.02.001

Stephens R, Wikstrom N (eds) (2000) Metropolitan government and governance: theoretical perspectives, empirical analysis, and the future. Polit Sci Quart (Oxford University Press) 115(1):164–166. https://www.jstor.org/stable/2658064

Tomas M (2020) Models of metropolitan governance. Ajumtament de Barcelona. Fundación para la Universitat Oberta de Catalunya, available at: https://www.metropolis.org/sites/default/files/resources/m1-_final.pdf

World Bank (2017) Casablanca municipal support program-for-results. International Bank for Reconstruction and Development Program, Appraisal Document. Available at: https://ewsdata.rightsindevelopment.org/files/documents/95/WB-P149995_L7GBVXG.pdf

World Population Review 2022. Available at https://worldpopulationreview.com/world-cities/casablanca-population

Zaki L (2019) Decentralization in Morocco: promising legal reforms with uncertain impact. In The Arab Initiative Reform, available at https://www.arab-reform.net/pdf/?pid=5958&plang=en

Chapter 10
International Standards: An Innovation in Urban Governance? The Case of Urban Development in Conakry

Pascal Rey and Margot Petitpierre

10.1   Introduction

At the end of the twentieth century, it was recognised that development projects financed by the World Bank could generate negative social and environmental impacts that outweighed their positive impacts. This was particularly true in the contexts of involuntary displacements and, more generally, cultural heritage. This observation led Michel Cernea to formulate an initial set of "guidelines" (Cernea 1988) that later became the World Bank Group's environmental and social standards and remain in force today. In 2003, the Equator Principles were established on the basis of these first sets of standards, and over the years, they have been ratified in dozens of countries around the world. As a result, the Equator Principles have become a benchmark for the implementation of public and private projects in developing countries when international funding is involved. Today, 38 countries recognise the Equator Principles through 129 public and private financial institutions. The Principles have continued to evolve, being last updated in 2018 to highlight the importance of the active participation of all stakeholders in the management of spatial planning and urban development. Here, we highlight issues related to their implementation in the context of infrastructural development in Conakry (Republic of Guinea) and reflect on the innovation that these standards represent in terms of governance, both in terms of project development and project implementation.

P. Rey (✉)
Institute for Social Research in Africa (IFSRA), Ouagadougou, Burkina Faso
e-mail: pascal.rey@insuco.com

M. Petitpierre
General Manager for Asia, Insuco, Phnom Penh, Cambodia
e-mail: margot.petitpierre@insuco.com

© The Author(s), under exclusive license to Springer Nature Switzerland AG 2024   151
K. Darmame, E. Ross (eds.), *Local Governance and Development in Africa and the Middle East*, Local and Urban Governance,
https://doi.org/10.1007/978-3-031-60657-1_10

Phases of urban development projects of national interest that are governed by the Equator Principles include consultation, the identification and assessment of socio-economic impacts (Dendena and Corsi 2015), and the relocation and resettlement of affected people. These principles provide a framework for assessing and accounting for environmental and social issues (Leduc and Raymond 2000); here, we focus on the social standards.

Based on concrete examples of differently financed infrastructure projects in Conakry, this article reflects on the advances and innovations represented by the increasingly widespread application of the Equator Principles to include local stakeholders in governance processes. Meanwhile, we consider the paradoxical fact that if these standards are not also applied to other projects in the same area, they can create pronounced social disparities.

To this end, we first present the national geographic and legal context of Conakry, then return to the normative international reference. In these two sections, we present examples that highlight the disparities generated by the varied consideration of stakeholders in urban development projects. Based on observations of the consequences of these disparities, we will probe how well the implementation of the Equator Principles truly represents progress. Finally, we will discuss emerging perspectives related to the evolution of governance practices in Conakry and their correlation with the new government's current and future choices of international partners.

This article is based on three different types of information sources: (1) case studies of the implementation of private and public projects in Conakry that were subject to international standards, (2) literature reviews, and (3) semi-directed consultations with people displaced by public projects not subject to international standards.

10.2 A Limited National Legislative Framework

In Conakry, the City Planning Code (*Code de l'urbanisme*), the Land Tenure Code (*Code Foncier et Domanial*; Rey 2011) and the Construction and Housing Code (*Code de la Construction et de l'Habitation*), all Guinean texts, are used as references for managing urban development. For projects declared to be in the public interest, the 2019 Environmental Code is the reference; generally speaking, it guides project promoters through the processes required to address environmental and social concerns.

The Guinean Agency of Environmental Evaluation (*Agence Guinéenne d'Evaluation Environnementale*, AGEE), under the supervision of the Ministry of the Environment and Sustainable Development (*Ministère de l'Environnement et du Développement Durable*, MEDD), is responsible for enforcing assessments of environmental and social impacts—whether positive or negative—that are required by the Environmental Code. Stakeholder information processes are a requirement of the country's legal and regulatory framework at all stages of project implementation.

In particular, the legal and regulatory framework requires that the results of environmental and social assessments be made available to stakeholders. This feedback must be provided both in the local study area and at the national level. To do this, the AGEE requires that 23 executives from the ministerial departments involved in the project in question form a Technical Committee for Environmental Analysis (*Comité Technique d'Analyse Environnementale*, CTAE).

In addition to the Environmental Code, the Land and State Code governs issues relating to displacement and resettlement processes. The Expropriation for Causes of Public Utility (*Expropriation pour cause d'utilité publique*, ECUP) procedure provides for the participation and consultation of people affected by a project. In April 2021, the MEDD amended its regulations to require the development of a Resettlement Action Plan (RAP) whenever a project has a physical or economic impact[1] on more than 50 people. A RAP must be created and validated by the AGEE. However, the practical procedures for implementing and validating population relocations are not specified in these new regulations. Furthermore, these regulations postdate, and thus do not apply to, the case studies cited herein.

In Guinea, the recognition of public utility follows validation of the public utility survey and establishes the fact that individual or collective property rights, as designated by the Guinean Constitution and the United Nations' Universal Declaration of Human Rights, are less important to the general public than the project in question. A declaration of public utility may also be issued by presidential decree for projects with major land or social impacts, or by order of the Ministry of Urban Planning, Housing and Territorial Development for projects of more limited impact.

The reality observed on the ground can be quite different from what one would expect based on this legal framework. Despite ideal geomorphologic and climatic conditions[2] and its numerous natural riches, Guinea remains one of the poorest countries in the world, ranked 178th in the United Nations' 2020 Human Development Index. Sadly, Conakry, Guinea's peninsular capital, well illustrates the general situation and problems of the country. Despite the country's natural wealth, its infrastructure and public services are poorly developed, and the lack of coherent governance is glaring. Various urban development plans for the city (*Plan de Développement Urbain de 1989, Grand Conakry 2040*) have not succeeded in building a capital with a clear master plan and, today, its inhabitants suffer as a result. For example, sanitation services are virtually non-existent, connections to electricity are unreliable, road and rail installations are impassable or insufficient,

---

[1] Physical displacements are cases in which housing and/or property losses due to land acquisitions by a project require the relocation of affected people to a new site. Economic displacements are losses of resources, income, or means of subsistence due to land acquisitions or restrictions of access to certain resources (land, water, forest) upon the construction or operation of a project or its ancillary facilities.

[2] The Republic of Guinea is a coastal West African country that benefits from soils and subsoils extremely rich in minerals (bauxite, iron, gold, diamond, etc.). Guinea receives sufficient rainfall that, combined with its favorable geological formations, gives it some of the most important hydroelectric potential in West Africa.

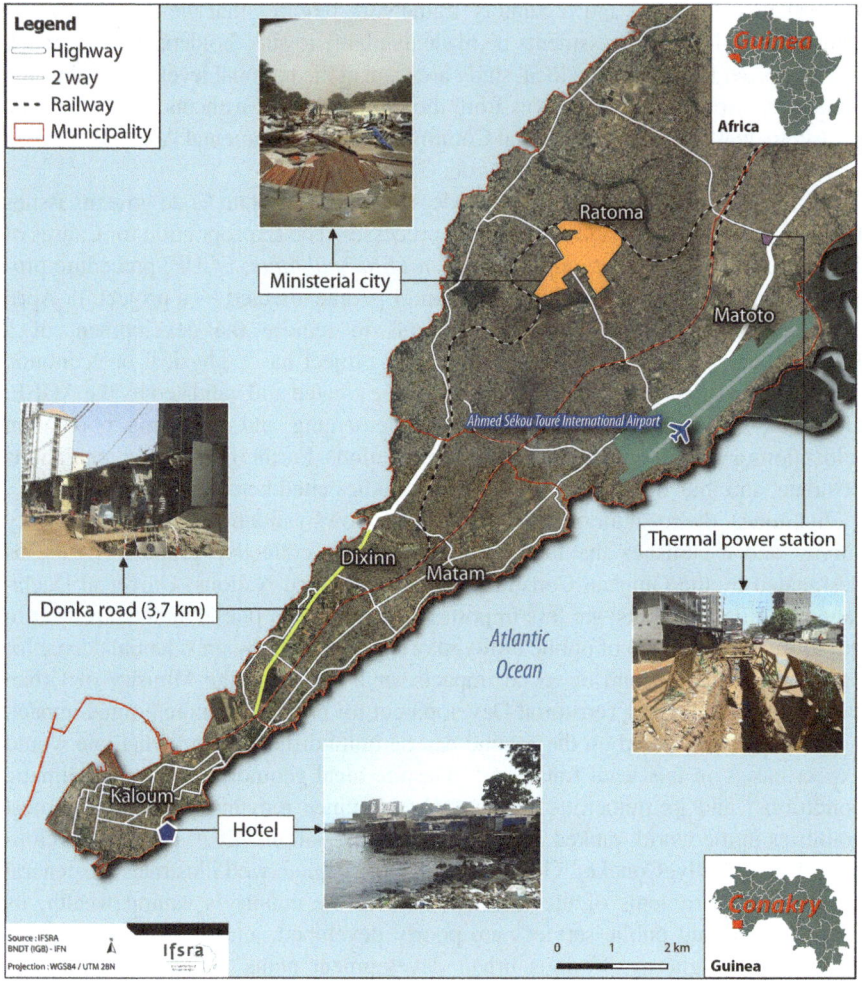

Fig. 10.1 Location of infrastructure projects in Conakry
Source: the authors

and health services are incompetent or even unsanitary. Nonetheless, with the aid of numerous public and private investors, the government continues to implement urban development projects aiming to improve the capital and the lives of its inhabitants and workers.

In this article, two infrastructure projects linked to the urban development of Conakry are taken as examples (Fig. 10.1). The first was initiated some twenty years ago and involves the construction of a ministerial housing estate in four of the city's districts: Kaporo-Rails, Kipé 2, Dimesse, and Dar Es Salam; this project was modified in 2019, but has not progressed at the time of this writing. The second is the widening of one of Conakry's main roads, the Route de Donka; under the aegis of a Sino-Guinean framework agreement, it targets a 3.7-km-long section of road (Rey and Mazalto 2020a) and falls under a wider project to build and rehabilitate a 70-km-long network of roads. We focus on these projects because they illustrate cases of financing that were not subject to compliance with the international normative framework, but only with the Guinean normative framework described above.

The ministerial city project is an attempt to deal with Conakry's unique geographic constraints, its situation at the tip of a peninsula, by reorganizing the flow of traffic in the city; it aims to avoid massive morning influxes to the tip of the peninsula and equally massive evening outfluxes to the suburbs, causing daunting traffic jams. To this end, the government of Lansana Conté (1984–2008) initiated a polycentric development policy in the early 2000s that included a project to build a new centre of attraction outside the city centre, in the Kaporo-Rails quarter. The project was then broadened to integrate foreign embassies, businesses, and other services.

Already in the early 2000s, numerous clashes during the eviction of occupants led to the intervention of the army and, ultimately, deaths (International Federation for Human Rights 2004). This policy of "liberating" areas continued under the presidency of Alpha Condé (2010–2021) and, in 2019, the NGO Human Rights Watch denounced the thousands of physical and economic displacements carried out under the pressure of bulldozers with the sole justification of a 1989 decree allocating the land in the area to the State (Human Rights Watch 2019). Alpha Condé recalled that *"seules les personnes titulaires de titres fonciers dûment authentifiés ont droit à une indemnisation de la part de l'Etat. "*[3] This demonstrates why so few Persons Affected by the Project (PAPs) were able to assert their rights. As will be discussed below, in addition to the low rate of securitisation of land in Guinea, it is normally not allowed to claim a private title to public land.

The Route de Donka is one of Conakry's main arteries, providing access to the city centre at the tip of the peninsular capital. For numerous reasons, the road was very congested, and even dangerous in places. As soon as the widening of the road began in October 2019, the project received a great deal of attention from Conakrians, particularly motorists and motorbike taxis. The project has gone through several stages including the destruction and reconstruction of the road in small sections and the introduction of alternating traffic patterns. The outbreak of the COVID-19 pandemic led to the suspension and delay of the work, and drains were left protruding from the road. Although the Guinean legal and regulatory framework requires the implementation of a stakeholder consultation process and the establishment of a relocation procedure, neither was initiated despite the severity of the project's impacts. The situation speaks for itself and can be seen at a glance: the removal of pavement has caused businesses run by women to disappear, and houses along the road have literally been cut in two, making the interiors of kitchens, bathrooms and living rooms visible to everyone living or circulating nearby. This is all despite the fact that the government must consult all stakeholders and provide

---

[3] Only those holding duly authenticated property deeds are entitled to compensation by the State.

compensation and/or indemnities previously presented and validated *via* a matrix of compensation rates and eligibilities in cases requiring displacement; that is, Environmental and Social Impact Assessments (ESIAs) must be conducted and People Affected by the Project (PAP) must be informed. This example highlights the limits of the government's capacity, through its institutions, to implement urban development projects of national interest that do not comply with international standards. Although the national legal and regulatory framework is extensive, it is difficult to guarantee compliance when the state is the main responsible party, regardless of any cooperation with an institution in compliance with international standards (Fortier 2014).

10.3 International Standards and Stakeholder Consideration

Of the projects that contribute to the urban infrastructural development of Conakry, public projects are placed under the supervision of the State but require funding from international donors, whereas private projects are financed by international banks subscribing to the Equator Principles and can be declared to be of public utility. Public and private projects alike are subject to compliance with international environmental and social standards.

The formulation and implementation of such standards followed the observation that development projects can generate negative externalities detriment to residents and their households, cultural heritage, and the general social fabric. In the 1970s, the *ex-ante* evaluation of development projects became widespread, and the World Bank adopted the practice in the 1980s, applying the methodology to its projects in developing countries and incorporating considerations of social impacts. The World Bank also developed the Environmental and Social Framework (ESF), a document that attained its current form, after numerous reviews, with the 2018 revision.

The ESF comprises ten Environmental and Social Standards (ESSs):

1. ESS 1: Assessment and Management of Environmental and Social Risks and Impacts. This standard guides the promoter, both generally in the consideration that they will encounter environmental and social issues during the implementation of their project, and more precisely at each major stage. This article is largely concerned with this first standard.
2. ESS 2: Labor and Working Conditions.
3. ESS 3: Resource Efficiency and Pollution Prevention and Management.
4. ESS 4: Community Health and Safety.
5. ESS 5: Land Acquisition, Restrictions on Land Use and Involuntary Resettlement. We make many references to this ESS herein, which guides the promoter in limiting involuntary displacements (when inevitable), adopting appropriate measures to minimize negative impacts on PAPs, and implementing appropriate planning and management practices.

6. ESS 6: Biodiversity Conservation and Sustainable Management of Living Natural Resources.
7. ESS 7: Indigenous Peoples/Sub-Saharan African Historically Underserved Traditional Local Communities.
8. ESS 8: Cultural Heritage.
9. ESS 9: Financial Intermediaries.
10. ESS 10: Stakeholder Engagement and Information Disclosure. This ESS is also a prominent theme throughout this article. It recognizes the importance of open and transparent consultations between the Borrower (promoter) and project stakeholders as an essential element of good international practice.

These standards apply to all projects financed by the World Bank, without restriction or exemption, and all States and international public donors are invited to comply.

The International Finance Corporation (IFC), an organization of the World Bank Group established in 1998 and dedicated to the private sector, adopted the World Bank model by developing its own eight Performance Standards on Environmental and Social Sustainability in 2006 (last revised in 2012). These standards target private investors supported by the organization (IFC), but banks, investment funds, and multinationals are widely encouraged to apply them as well.

These two normative corpuses are intended to supplement national legal and regulatory frameworks when they are insufficient to control environmental and social risks. Public and private beneficiaries of grants and loans from the World Bank/IFC, donors, and banks that have ratified the Equator Principles are contractually committed to respecting these standards.

Through its Operational Manual and Environmental and Social Framework, the World Bank (and the IFC through its Performance Standards) quickly and widely imposed the *ex-ante* evaluation of public and private development projects in developing countries. Indeed, a series of alignments with these standards on the part of investment banks (Equator Principles, 2003) and development aid agencies (Paris Declaration, 2005) was observed. For their part, developing countries have progressively imposed *ex-ante* evaluations and environmental permitting, referencing (more or less) the World Bank standards.

It is therefore the development finance sector that is little by little taking on the role of regulator. To be financed, an operator (public or private) must manage the environmental and social risks of their project. They must demonstrate that they have studied the area in which the infrastructure will be installed, evaluated the environmental and social risks, planned measures for their avoidance, mitigation, and compensation, and previsioned to ensure continuous monitoring and evaluation. To these ends, public and private funders require specific documentation. ESIAs must include a stakeholder consultation plan, environmental and social baseline studies, an impact assessment, and an Environmental and Social Management Plan. The latter may be accompanied by a PAP in cases of physical population displacements (impacting residences) and a Livelihood Restoration Plan in cases of economic displacements (impacting sources of revenue or subsistence). Other assessments and plans more specific to the project characteristics and impact area may also be required.

As early as the project planning phase, all stakeholders must be consulted to account for their expectations, needs, and fears. This requirement is governed by the first guidelines of international environmental and social standards relating to project feasibility. The aim is to reduce potential negative effects, reinforce positive effects, and ensure that all stakeholders are informed and willing to express their opinions. In particular, this involves addressing the residents and beneficiaries of the project.

The standards then set requirements for the preparation of impact assessments. All standard assessment topics are considered: culture, heritage, the economy, and, above all, the stakeholders because the basic social survey within the impact study targets their socio-economic status. Impact assessments then lead to the development of an Environmental and Social Management Plan (ESMP). From a social point of view, an ESMP must ensure the social integration of the project in its area (i.e., the neighbourhood or town, depending on the scale of the development project and its impacts). From an environmental point of view, it must ensure general compliance with biological and physical requirements.

The World Bank's international standard that received the most updates in the 2017–2018 revision was ESS 10, which:

[. . .] recognizes the importance of open and transparent engagement between the Borrower [promoter] and project stakeholders as an essential element of good international practice. Effective stakeholder engagement can improve the environmental and social sustainability of projects, enhance project acceptance, and make a significant contribution to successful project design and implementation. (World Bank 2017:97)

This new form implies that all types of projects should systematically include a stakeholder engagement plan, expressing a true desire to include all stakeholders in urban governance.

International standards therefore require that those affected by a project are systematically involved in all phases of the project once its feasibility has been established. The Stakeholder Engagement Plan (SEP) includes an information channel for consultations to address the expectations, fears, and needs of each stakeholder group. A perfect illustration of this reality is a project to expand a hotel complex in Conakry's city centre that was blocked by the stakeholders: although backed by international private funding, supported by the presidency, and recognised as being of public utility, the project was denounced by residents who had illegally settled the area (at the time considered to be squatters). The absence of any explicit requirement in Guinea's legal and regulatory framework to draw up such SEPs,[4] even though

---

[4]Neither declaration A/2013/474/MEEF/CAB of 11 March 2013, adopting the general environmental evaluation guide, nor 1/2022/1646/MEDD/CAB/SGG of 25 July 2022, establishing the administrative procedures for environmental evaluations, explicitly reference a stakeholder engagement plan. Nonetheless, in practice, the AGEE in charge of evaluating ESIAs requires that a stakeholder engagement plan be included in the submitted documentation.

public consultation is provided for, is one of the biggest discrepancies with international standards.

An ESIA is required by both normative international references. Notably, an ESIA must include a baseline study and, informed by the results and consultations, an impact evaluation must be performed to develop a plan to integrate all measures to mitigate negative impacts. As mentioned above and as underlined by Rey and Petitpierre (2022), when the State is independent of international financiers in promoting a project, these studies are rarely performed, at least not at the quality level necessary for such an exercise. The State does not always conform to all the legal prerequisites to implement its projects, and acquiring construction or exploitation permits generally follows a simplified process, mostly via negotiations among implicated Ministries.

However, ESIAs/baseline studies are important in terms of project management, particularly because they allow stakeholders, especially PAPs and residents, to contribute to their definition. The case of the construction of a thermal power plant in the Matoto district illustrates this. Following an impact study, the project promoter drew up an ESMP to limit negative impacts on the population. This ESMP accounted for the expectations of local residents, particularly with regard to supporting them during the construction period. For example, an informal vendor received compensation equivalent to 2 months' business for having his stand moved by 200 m during the construction period. The promoter not only compensated the vendor for potential loss of income, but also helped move their stand and, through five consultations during the construction period, ensured that the compensation received was sufficient to offset any loss of income.

In Guinea, the disparity in the consideration of PAPs between the national legislation and international standards in terms of involuntary urban displacements speaks for itself. There are forced displacements with only a semblance of consideration for landowners and accountability towards all those affected, whether economically or physically, periodically or permanently, regardless of legal statute. In contrast, the World Bank reference takes equal equally account of people whose occupation of a project site may be irregular, on the condition that they occupied the land before a fixed eligibility date (ESS 5, paragraph 34.c).

10.4 International Standards: An Innovation in Urban Governance

The Equator Principles represent true progress in urban governance for numerous developing countries. In terms of social aspects, they essentially rely on taking into account all stakeholders and placing part of the governance of urban development in the hands of citizens and land users. Why respect for these standards is guaranteed neither by national nor international authorities, it is leveraged through financial instruments; should a promoter not respect this normative framework, they will foreclose any future funding, be it public or private. Respect of these standards is thus better guaranteed by financial leverage, as gauged by financiers' refusals to finance projects, or by their insistence that the full complement of consultations or assessments be performed to secure financing. There are thus two levels to reinforcing the idea that stakeholders be included in governing projects: the written rules, and how effectively they are taken into consideration. In cases where only the State is implicated in project management, this last point represents a veritable innovation in terms of respect for national legislation (Olivier de Sardan 2004).

Although the international environmental and social standards are not a toolbox in themselves, they represent a set of directives and positive changes (Rey and Mazalto 2020b). For example, frameworks for stakeholder consultations consider their expectations and wishes, and evaluate the impacts that may weigh on the project design. Most of all though, the fair consideration of all persons affected by a project of general interest represents considerable progress in terms of respecting human rights, which should become more generalized.

Finally, these standards also represent an innovation by providing valuable support to States whose legislative corpus is under construction or reform. Indeed, national legislative codes often refer to these standards for questions related to social or environmental issues. This supports the financial leverage with legal leverage, which is essential for obtaining permits for project implementation and operation. For example, a 2014 decree on ESIAs (although targeting the mining sector)[5] requires the promoter to comply with national standards and, failing that, with IFC standards and/or the Equator Principles.

However, the expectations of a person concerned by an urban development project might be taken into account and treated differently depending on the type of funding acquired for the project in question. Revisiting the example of the project to build the ministerial city of Kaporo-Rails, landowners in possession of title deeds were not necessarily judged on the value of their investments. We spoke to Mr. S., a former resident of the Kaporo-Rails district. Mr. S. was in possession of a donation certificate for two plots of land acquired by his father before the proclamation allocating the land to the State in 1989, as well as an ownership deed dated to 1989. Nonetheless, he was displaced without receiving any compensation. Conversely, some of his neighbours whose homes were also demolished received compensation, although the amount was set arbitrarily and never exceeded the equivalent of €2000. Such arbitrary compensation practices neither comply with international standards nor correspond to the recommendations of the national legislation. In fact, it is required that the compensation offered be calculated based on an assessment of the level of damage.

---

[5] Decree D/2014/014/PRG/SGG of 12 January 2014 adopts a guideline for carrying out an ESIA of mining operations, taken in application of the Environment Code and the Mining Code, and inspired by the general guide to ESIAs adopted by Order A/2013/474/MEEF/CAB of 11 March 2013.

Title deeds to spaces reserved for the public cannot be recognised *a priori* by the State. By definition, they relate to areas that cannot be privatised, and, although granted by the government itself, in principle they cannot be used privately. In addition to proposing unrealistic and extremely low compensation rates for the assets to be compensated (and that do not comply with international standards), the State also fails to recognise the reprehensible practices of its own institutions, even though it stands to benefit from them. Affected people then have no choice but to accept the compensation offered, however low it may be, and leave.

This example is a perfect contrast to that of the thermal power station and the informal vendor, and a major paradox thus arises. On one hand, an informal vendor not recognised by the State was taken into account through consultations, affected by a temporary economic displacement (moving their stand by 200 m), and received compensation of around €1500. On the other hand, an individual able to present a title deed recognised by the Land and Property Code was affected by a permanent physical displacement (forced expropriation no less) but did not receive compensation because the State refused to recognize a title to public land. This paradox is compounded by the fact that, in the case of the owner, such titles were often sold by prior administrations for the personal benefit of a corrupt civil servant.

In the cases described in this chapter, international standards could only make a limited contribution to urban management. This raises the question of how to standardise displacement issues relating to projects considered to be of national interest. States fear that the implementation of the Equator Principles will lead to an increase in spending on infrastructure projects. Although the legitimacy of States' concerns in this respect may be questioned, it is all the more interesting to highlight the fact that legal and regulatory frameworks increasingly refer to international standards, demonstrating a definite move towards best practice and suggesting the increased integration of the populations concerned by urban governance.

10.5 Political Choices and Perspectives for Conakry

The political events that have occurred in Guinea over the past 15 years require that we reflect on perspectives for developing Conakry and discuss the factors that have blocked the implementation of the policies initiated in the early 2000s. Indeed, for to a casual observer, the effects of the political decision to transform Conakry into a polycentric city are not obvious. In 2022, Conakry seems to be a perpetual construction site, with the completion deadlines for wok delayed time and again. Many quarters have also seen the start of projects, notably those related to sanitation, suspended. Although traffic in the has city never flowed smoothly, in recent years it has become increasingly jammed due to the plethora infrastructure projects of various magnitudes.

The works in question are largely carried out by Chinese companies, and it is easy to place this in the context of the accords signed between China and Guinea in 2017. On 6 September 2017, China and the government of Alpha Condé signed a framework agreement granting Guinean mineral resources to Chinese companies in return for up to $20 billion US worth of infrastructure financing (Bouessel du Bourg 2017). The results of this investment in infrastructure development are difficult to see, whether in the capital or elsewhere in the country. Indeed, Chinese enterprises straddling the line between private and public (Rey and Mazalto 2020a) are not necessarily required to respect environmental and social standards because their funds often come from public Chinese banks. A priori, they therefore represent a risky form of partnership in terms of land development and environmental, physical, and biological impacts. As such, the projects resulting from this Sino-Guinean agreement do not follow the normative reference described here. They rarely involve stakeholders in their development and rarely consult and compensate PAPs.

The amount of consideration accorded urban residents, or citizens in general, can therefore depend less on the government in place than on the international partnerships into which it enters. The latest events in Guinea, i.e., the rise to power of Col. Mamady Doumbouya and his future partnership choices, must be scrutinized in this context; it is the international partnership decisions that will define the future of Conakrians for years to come.

10.6   Conclusions

In the absence of a supranational entity capable of exercising influence over States to compensate for any shortcomings in national legislations in terms of providing for stakeholders in the context of urban development, the latest updates to the environmental and social standards of the World Bank Group and the many financial institutions that have ratified them represent a real contribution to modes of urban governance. This financial leverage, in the absence of adequate national legislation, is certainly the most important for ensuring that citizens are truly considered in the development of their own city.

The normative international framework and its latest reforms emphasize stakeholder engagement. They include the consideration of PAPs but can create considerable inequalities in stakeholder treatment. The source of funding, which reflects how likely the international standards are to be applied, is a determining factor in the creation of these inequalities. Residents impacted by projects are faced with a kind of lottery and their fates depend on factors beyond their control. As hard as it is to hear that an element promoting the better consideration of human beings can be responsible for such disparities due to the State's lack of consideration of its own citizens, international standards can be perceived as responsible for creating a social sieve (Rey and Petitpierre 2022).

Nonetheless, the trend is moving in a positive direction, i.e., towards adequately integrating projects into their social environments. The incorporation of these international standards into more and more projects and progressively into national legislative corpuses and the habits of city dwellers and users, i.e., in terms of their involvement in the development of their city, gives hope that this new form of governing development projects will expand and progressively promote citizen participation in urban governance. It is thus the widespread application of international standards that can avoid the inequalities in stakeholder treatment described above.

Although this increasing tendency to apply international standards seems irreversible, its acceleration and generalization remains dependent on the political choices that will be made in terms of international partnerships. The inclusion of actors not subject to the application of international environmental and social standards will slow, rather than accelerate, the involvement of citizens in the policies put in place to develop Conakry.

References

Bouessel du Bourg C (2017) Mines: la Chine et la Guinée signent un accord à 20 milliards de dollars. Jeune Afrique 8 Septembre 2017 https://www.jeuneafrique.com/472655/economie/mines-la-chine-et-la-guinee-signent-un-accord-a-20-milliards-de-dollars/. Accessed 21Sept 2022

Cernea M (1988) Nongovernmental organizations and local development. World Bank discussion paper (40)

Dendena B, Corsi S (2015) The Environmental and Social Impact Assessment (ESIA): a further step towards an integrated assessment process. J Clean Prod 13:1–13

FIDH (2004) Mission Internationale d'Enquête, Guinée, une démocratie virtuelle, un avenir incertain. FIDH, Conakry

Fortier J-F (2014) La déterritorialisation en tant que désinstitutionalisation de l'espace politique. Réflexions sur la gouvernance territoriale et les conditions de possibilité d'une critique sociale. In: Soulière M et al (eds) Visages contemporains de la critique sociale, réflexions croisées sur la résistance quotidienne. Éditions ACSALF, Montréal, pp 107–126

Human Right Watch (2019). https://www.hrw.org/fr/news/2019/06/18/guinee-des-expulsions-forcees-draconiennes. Accessed 20 Sept 2022

Leduc G, Raymond M (2000) L'évaluation des impacts environnementaux: Un outil d'aide à la décision. Éditions MultiMondes, Sainte-Foy

Olivier de Sardan J-P (2004) État, bureaucratie et gouvernance en Afrique de l'Ouest francophone: Un diagnostic empirique, une perspective historique. Politique africaine 96:139–162

Rey P (2011) Droit foncier, quelles perspectives pour la Guinée ? Réflexion sur la réforme foncière à partir de l'exemple de la Guinée Maritime. Annales de géographie 679:298–319

Rey P, Mazalto M (2020a) Quand le développement des territoires miniers brouille les frontières entre les secteurs public et privé: Cas du secteur minier en Afrique de l'Ouest. Mondes en Développement 189:81–97

Rey P, Mazalto M (2020b) Reconciling standards and the operational needs of mining projects in Africa: examples from Guinea. Extr Ind Soc 8(1):23–31

Rey P, Petitpierre M (2022) First- and second-class citizens? Governance disparity in the urban development of Conakry, Republic of Guinea. Reg Dev Reg Policy 60:21–31

World Bank (2017) Environmental and social framework. World Bank Group editions, Washington DC

Part III
Public Services, Housing and Public Space

Chapter 11
Housing Provision in Poor Communities in Ghana, the Role of Non-State Actors

Esther Yeboah Danso-Wiredu

11.1 Introduction

This study looks at community-based associations and their effects on general housing provisions in selected communities in Ghana. To achieve sustainable urban development, it is important growth and development are measured at the local levels. This is because the SDGs, especially the SDG 6 of clean water and sanitation, 10 of reduced inequality and 11 of sustainable cities and communities cannot be achieved if development does not start from the local level, usually in partnership with other levels of governance as it is clearly stated in the first paragraph of the SDG-17 that 'A successful development agenda requires inclusive partnerships—at the global, regional, national and local levels—built upon principles and values, and upon a shared vision and shared goals, placing people and the planet at the centre (https://www.un.org/sustainabledevelopment/globalpartnerships/).

A major characteristic of growth in African cities is the impoverished nature of this growth. This feature is common in the developing countries, and Sub-Saharan Africa is the most affected (Moreno and Head 2011). One way to measure urban growth and development is through urban housing. For lack of affordability in formal housing markets, the urban poor find themselves living in vulnerable areas within cities. Houses, infrastructure and amenities are mostly in deplorable state in these areas. The urban poor are usually forced to find shelter in informal structures located in a variety of places (Serageldin et al. 2003). Despite the inhumane conditions in which they live, residents of poor urban areas make meaningful contributions to urban life. They provide low-cost housing for immigrants arriving from the rural areas, and to host them till their eventual absorption into urban life

E. Y. Danso-Wiredu (✉)
Department of Geography Education, University of Education, Winneba, Ghana
e-mail: eydwiredu@uew.edu.gh

(Neuwirth 2007). There exist in poor communities a variety of complex activities to make a living (De Boeck et al. 2010), though some are life threatening. How are they able to do this, in most case without the assistance of the state? They do this through participatory approach.

Ghana's urban population has been increasing from 7.8% of the country's population in 1921 to 50.9% in 2010. And 54.68 percent of Ghanaians lived in urban areas as at 2016 (GSS 2017). The number of people living in urban areas has exceeded the number living in rural areas since 2008 (Ghana Statistical Service 2012). The growth is explained by both natural growth as well as rural-urban migration. Rural-urban migration is a major phenomenon in Ghana, especially among the youth (Van Der Geest 2011; Cassiman 2008, 2010; Opare 2003). Migrants look for available space to reside in the city. Such places are usually not safe for human habitation. Also found in most cities in Ghana are indigenous communities which have deteriorated for lack of maintenance of both houses and services.

Ghana struggles with urban poverty and this physically manifests itself in the conditions of housing and community infrastructure in parts of the country's major cities. Ghana faces enormous challenges with housing provision for its lower income groups. Houses built by the private sector are expensive and low-income population cannot afford them, hence, the explosion of poor communities in the cities.

The research highlights the different strategies low-income communities in Accra and Winneba employ to access general housing facilities depending largely on non-state governance systems. The study tries to answer the question, to what extent do non-state governance structures regulate access to general housing resources in the study areas?

11.2   Social Capital, Community Based Associations and Participatory Development at the Local Level

The centralization of development in the form of top-down approach has been challenged as prohibiting smooth development at the local level. Participatory development is the involvement of the marginalised groups in society in decision making which affect their lives. Stakeholders' involvement in local development is the preferred choice of development (Mohan and Stokke 2000). This kind of development is also embedded in its participation and empowerment at the local level. And it accepts the collective mobilization of marginalised groups, believing that power is in the hands of individual members of the society for collective usage (Mohan and Stokke 2000). It is argued that development without participation from the grassroot level is unsustainable (Cornwall 2003). Participatory development includes both the powerful and less privilege in the society, it has therefore become necessary in policy planning globally (Cornwall 2003). It encourages all people to participate in a political process which concerns them in forms that will help them gain tangible benefits (Chambers and Conway 1992).

It is important that community participation in developments stems from a stable group of people who can identify themselves as members of the group based on similar norms, backgrounds, priorities, aspirations and interests (Anyidoho 2010). And that these groups of people participate in development projects in ways that benefits them. Here, communities of place are combined with communities of interests (Cornwall 2003). This is the reason why the study looks at local participatory development from the lenses of a well-established social relationships and the benefits thereof (social capital).

Social capital can be explained as the networks of relationships among community members, thus, people who live, work and perform other common activities in the same geographical area, be it virtual or physical, where their relationships result in a better functioning of their community. Putnam defines social capital as features of social life-networks, norms and trust that enable participants to act together more effectively to pursue shared objectives (Putnam 1995:664). From Putnam's perspective, social capital enhances reciprocity and increases the speed at which information flows for mutual benefits and trust among community members (Mohan and Stokke 2000). Social capital is central to this study because local associations which form the focal point of this research are strongly linked to varying forms of social relationships leading to social capital. The forms of social relationships discuss in this study become the basis of trust on which people totally rely. This study looks more at the communal social capital and its effects on housing provision because in the study communities most decisions relating to means of subsistence are made collectively and the effects are also collectively felt.

Community based associations play important role in the livelihoods of the study population. Most of them are seen as informal in nature. Informal institutions are socially shared rules, it can also be established aspect of a society, usually unwritten, that are created, communicated and enforced (Helmke and Levitsky 2012). The institutions or rules that govern informal governance are hardly written (Hyden 2008; Messer and Townsley 2003; Helmke and Levitsky 2012).

Informal institutions can simply be seen as local structured social systems which Stinchcombe and March (1965) describe as variables which form stable characteristics of a society outside organizations. In the study communities, institutions like kinship relations, societal norms and taboos play important roles in the everyday lives of the population and greatly affect their livelihoods. The rationality of informal organisations is embedded in cultural and social norms accepted by the members involved. The informal organisations and institutions basically constitute the non-state actors who form the governance structures in the study communities.

Another argument develop by this chapter is how best do we make local development in poor communities the responsibility of the state and non-state actors. Hence the study argues that community participation in local development through social capital is not enough, hence the need to go beyond the participatory approach to policy making which is possible with the involvement of the state.

11.3   Research Setting and Methodology

A community-based study was employed for the study. A community is described as a geographically delineated unit within a larger society (Berg et al. 2004:233). Communities selected as case studies are small enough to have a cultural homogeneity. There are interactions among members and they produce social identification within them (Berg et al. 2004). The study communities were selected based on typologies of poor communities in Ghana (Table 11.1). Old Fadama (Accra) represents a migrant-squatter poor community whilst Agbogbloshie (Accra) represents a poor community comprising both migrants and indigenous population. Penkye in Winneba represents an indigenous poor community in a small city and Akosua village also in Winneba represents a migrant-squatter poor community in a small city. The residents in the two communities in Accra are mostly people working in the informal sector and are dependent on the Agbogbloshie market and nearby markets for survival. The communities in Winneba on the other hand, are fishing communities and the residents work in the fishing or fishing related industry. The four communities are characterized as poor communities because their houses are mostly in deplorable states, the rooms are congested with people and things, the houses lack household amenities and the communities lack infrastructural development.

11.3.1   The Study Areas: Accra and Winneba

The Greater Accra Region is the smallest of the ten administrative regions in terms of land area occupying a total land surface of 3245 km^2, thus 1.4% of the total land area of Ghana (Fig. 11.1). The region has a coastline of approximately 225 kilometres; the vegetation is mainly coastal savannah shrubs. It is within the dry coastal equatorial climatic zone (Government of Ghana 2014). Accra is the hub of economic

Table 11.1 Categories of poor communities studied

Source: Danso-Wiredu, 2016

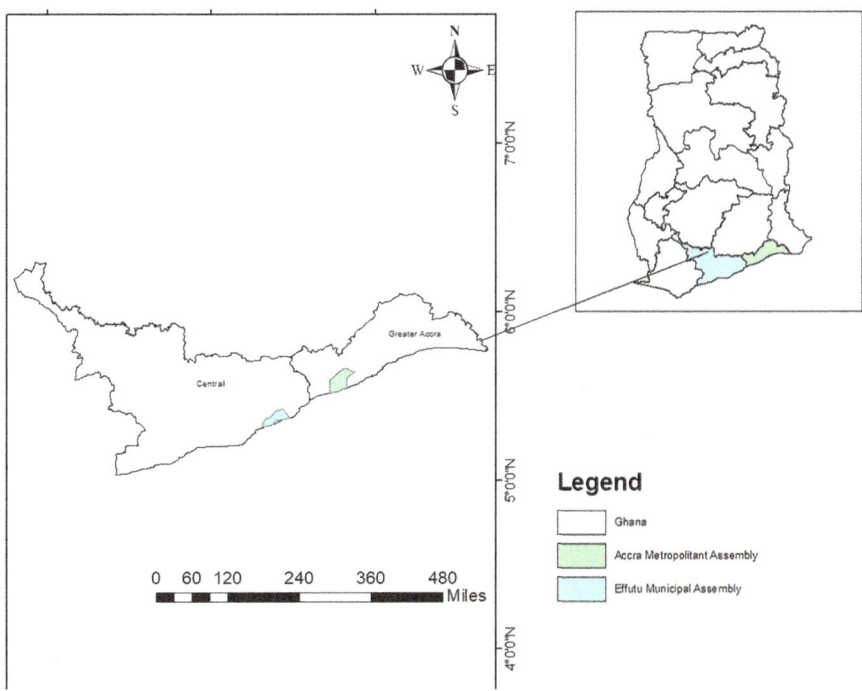

Fig. 11.1 Map of Ghana showing study areas
EMA Effutu municipal assembly, *AMA* Accra metropolitan assembly. Source: author

life of Ghana. Major manufacturing activities are located in the region; it is a marketing centre for both the formal and informal trading activities; it is also an important area for the country's financial, insurance, transportation and tourism industry. Fishing is an important economic activity for the people along the coast in the indigenous areas.

In terms of population, it is the most populated region with a population of 5,446,237 in 2021 by the National Population Census. (Ghana Statistical Service 2021). Greater Accra is the most densely populated region with a density of approximately 1300 persons per square kilometer (Ghana Statistical Service 2021). The Accra metropolitan area (AMA) is the most populated and developed of all the districts. Accra has been Ghana's capital since 1877 is in this district. The population increase is attributed to both natural growth and migration. About 44% of AMA population are believed to be migrants (Government of Ghana 2014).

Winneba is in the Central Region. The region occupies an area of 9826 km^2–4.1% of Ghana's land area. The region lies within the dry equatorial zone and moist semi-equatorial zone (Government of Ghana 2014). It has a population of 2201,83 (Ghana Statistical Service 2012) and 44,254 in 2021 (https://worldpopulationreview.com/countries/cities/ghana). It is the second most densely populated region after Greater

Accra with a population density of 224 persons per km². Winneba town is the capital of the Effutu Municipal Assembly. Winneba, like many other cities and towns in Ghana, has gradually expanded from a simple fishing community to a city. The main economic activity among the indigenous population is fishing. In the coastal communities, fishing industry employs more than 80% of the population. Other economic activities in the town include commerce, service and manufacturing (Effutu Municipal 2019). The service sector employs a large number of the migrants.

Old Fadama Old Fadama occupies 31.3 hectares of land along the Odaw River and the Korle Lagoon (Farouk and Owusu 2012; Danso-Wiredu 2016). A population census conducted in 2007 puts the number at 79,684, and a population density of 2424.18 persons per hectare (Farouk and Owusu 2012). The secretary of OFADA in January, 2013 stated that the population was over 100,000 (Danso-Wiredu 2016). And another estimation by another community leader in July 2021 puts the population at over 105, 000. The head count in 2007 revealed majority of households (65.9%) come from the three Northern regions and almost every ethnic group in the country with some coming from neighbouring West African countries live in the community. The population ethnic structure has not changed much since the head count.

Agbogbloshie The estimated population density of the residential part of Agbogbloshie alone, excluding the market area, is 1111.41 persons per hectare.[1] The community is made up of Gas and migrant population. Ga is the dominant language spoken at the residential part followed by Akan. But at the market, Akan is more spoken than Ga. A native Ga resident in an interview said majority of the people are Gas, Kwawus and Northerners. Ghana Statistical Service puts the Akans as majority followed by Gas (Ghana Statistical Service 2012). There are also migrants from different parts of Ghana and Africa living in the community. The community leaders estimate the population at the residential area to be over 20,000 in 2014 (Danso-Wiredu 2016) and over 22,000 in 2021.

Penkye and Akosua Village Both Penkye and Akosua Village are coastal communities in Winneba. They are both fishing communities and the majority of the population are employed in the fishing or fishing related industry. Penkye is an indigenous poor community; residents are indigenes of the area but residents of Akosua Village are entirely migrants from the Volta Region of Ghana. The Assemblyman of Penkye estimated, the population to be over 7000 in 2019. Akosua Village is a thin wedge shape land lying between the sea and the Muni Lagoon. The village is divided into three zones, Emma's village (zone1), Torgbui Akpadi (zone 2) and Gonyo's village (zone 3). The population is estimated to be over 1,200 by the residents in 2019.

---

[1] The population density was computed based on community leaders estimation of the population. The land area was also estimated on google earth.

11.3.2   Sampling Techniques and Size

The study employed the purposive stratified sampling technique. It is a sample within samples of particular units that vary according to key dimensions (Patton 2002). It is a hybrid approach which aims to select groups that show variations on a particular phenomenon but each is homogenous so as to compare the sub groups (Ritchie et al. 2013:79).

In total, a hundred and thirty-five people (135) were interviewed in the four communities. Many stakeholders and residents were interviewed in Old Fadama. They include ordinary residents, ethnic chiefs, organizational and community leaders. Based on information given by the secretary of the Old Fadama Association (OFADA) that the community has been divided into five zones A-E, stratified purposive sampling was employed to select respondents from each of the subgroups. Within each zone, respondents were selected based on their occupation so as to get as variant information as possible. Agbogbloshie was divided into two for the research: the residential part and the market and respondents selected from each category for the study. In Penkye, various fishing company owners, organizational members and leaders as well as residents were selected for the interviews. Akosua Village is divided into three zones, so stratified purposive sampling was used to select respondents from each zone for the interviews. Opinion leaders and organizational members and leaders in the community were also interviewed. Details of the sampled populations are shown in Table 11.2. Semi-structured interview guide and observation check-list were used for most of the interviews which were held with the aim of collecting information revolving around the main themes of the research: urban poor communities, housing and informal governance structures.

11.4   Structured Social Systems in the Communities

Different forms of social relations are identified in the study communities. Residents get help from people they usually have close ties with. The interviews reveal most migrants end up in Old Fadama, Agbogbloshie and Akosua Village because of invitation from relatives or friends who are already settled there. Most migrant therefore, have similar ties and relationships as exists in indigenous communities in the country. The basis of most organizations found in the communities stems from similar beliefs in the established institutional rules. There are also forms of social relations people derive by getting acquainted with community members or fellow traders. This might not necessarily be belonging to the same association, but friendship develops as a result of working together or living close to each other. The commonest social system apart from the family ones in Agbogbloshie, for example, is how traders in close proximity become friends and develop strategies to help each other.

Table 11.2 Demographic details of respondents

Respondents' details	Category	Old Fadama	Agbogbloshie	Penkye	Akosua village	Total
Gender	Male	33	8	16	14	71
	Female	24	16	13	11	64
Age group	−20	2	–	–	–	2
	21–30	20	2	7	3	32
	31–40	22	5	6	4	37
	41+	14	17	11	17	59
Marital status	Single	9	2	2	2	15
	Married	38	16	25	17	96
	Widowed	2	–	1	2	5
	Divorced	3	3	–	3	9
Level of education	No education	17	8	11	6	42
	Primary	17	5	9	9	40
	Junior/Middle	13	8	6	8	35
	Secondary/ vocational	7	1	1	2	11
	Tertiary	3	2	1	–	6

Source: See also Danso-Wiredu 2016

In Penkye and Akosua Village, there are strong community ties which extend to almost every livelihood acquisition, especially in accommodation and the fishing business. Parents pass on their rooms, equipment and jobs to their children. It is difficult for people who are not Effutu indigenes or Ewes to find accommodation in Penkye and Akosua Village respectively. The residents speak the same languages, belong to the same ethnic groups and have the same beliefs and traditions which guard their way of life, which includes the strong presence of the traditional belief system where some of them consult fetish priests. In all the communities, dispute arbitration by community leaders is common among residents. The trust in such centres stem from the belief in the established institutions. In Penkye, the arbitration system is well-structured in the form of traditional courts which settle matters of all kinds, from marital disputes to disputes among neighbours and business partners.

11.4.1 Organisational Types in the Communities

When people live together to share household facilities like toilet, showers and kitchen, sharing rooms in the same house or trading goods at the same market increase the bond of relationships among community members. These relationships are the lubricants for organizational formation and community development. The organisations are in various forms with different aims including religious, economic, ethnic, community development and welfare. There are criteria for members to be accepted or excluded in the community organisations. The various ways to become a

Fig. 11.2 Religious
affiliation of respondents

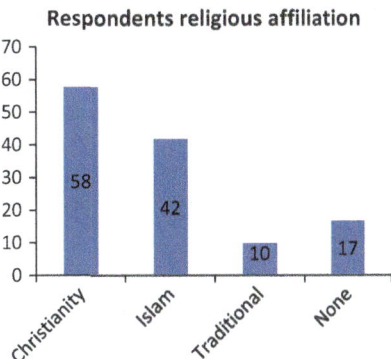

Fig. 11.3 Number of
organisations respondents
belong to

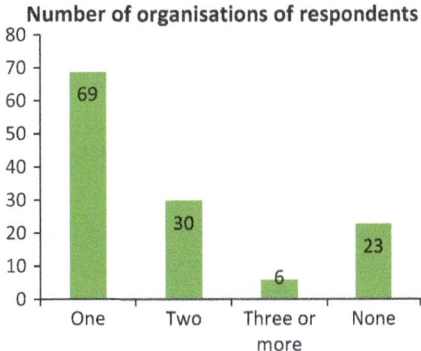

leader include appointments, elections, self-imposed and hereditary. The study
reveals almost every respondent belongs to a religious group (Fig. 11.2). Apart
from the religious organizations, most respondents belong to at least one of the
organizations discussed below (Fig. 11.3).

Ethnic Associations in Old Fadama Due to the diversity of migrants living in Old
Fadama, the community leaders decided to institute a chieftaincy system similar to
what prevails in the indigenous communities (See also Danso-Wiredu 2016, 2018).
This system is based on ethnicity, thus, residents from the same ethnic group form an
association and select one leader as chief of the association. This, according to them,
helps the community to take control over the management of the diverse population
with ease. Old Fadama is the only community among the study communities that has
this system. The chiefs play an important role in negotiations with the state and
NGOs on matters concerning Old Fadama. They also double as leaders of the various
ethnic associations in the community.

Economic Associations in the Communities There are associations open to people who are involved in similar economic activities (See also Danso-Wiredu 2016,
2018). These associations serve as the mouthpiece of the group members in negotiating with the state and the community leaders on their activities. They also manage

their members and mostly agree on similar pricing system and good conduct of members. Apart from their core mandates, the economic associations also attach welfare programmes and spatial development of their immediate environment to their tasks. They have strict rules of operation. Most of the leaders are voted by their members and they, therefore, have the mandate to sanction members who breach the rules they set. They are also responsible leaders who are usually fair because they could easily lose the trust of their members and lose their positions if they are not trustworthy.

Economic associations are found in all the study communities. In Agbogbloshie, traders automatically become members of the associations of their traded commodity. The associations see to the general welfare of their members, as well as the improvement and management of the market space and activities. There are numerous fishing companies in Penkye and a few in Akosua Village. Fishermen who work on the same canoe belong to the same company. The company names are derived from the names on the boats. Company members could be close or extended family members or friends. Companies have rules which regulate all members. The rules guide them in settling disputes if they arise and in sharing their proceeds. In Old Fadama, economic associations include showers and toilets operators, Kayayei, yam traders and Kokomba barbers.

Welfare Associations in the Communities In addition to the various welfare activities common to most local associations, all four communities harbor welfare associations. The welfare associations have grown organically from shared social linkage like those of traders who ply the same wares or operate in the same market, or residents from the same ethnic group. The aim of the welfare associations is generally to support their members financially in difficult times or in merry-making moments which involve expenditure of large sums of money. The benefits, in a way, serve as investments for their members. The criteria for leadership selection are different but mostly they are self-appointed leaders. They also operate within defined rules with sanctions and benefits which are known and understood by the members. Members of such associations pay dues on equal matching basis, thus, every member, including the leaders, pay the same. Benefits are also paid on equal matching criteria known to all members and leaders. The direct benefits from the group are the main reasons why people join, but there are indirect ones like developing acquaintances especially, among the migrants for future benefits. Most of the welfare associations also add leisure and sporting activities to their schedules, especially on weekends.

Local associational rules regulate means of access to general housing in the study communities. Strategies employed in accessing general housing are outlined and discussed in subsequent sections.

11.5   Access to General Housing Facilities in the Study Communities

Housing is a complex concept which has many dimensions. It includes access to and security of land tenure, shelter, as well as a number of facilities that often times come with them, such as toilets, showers and cooking facilities. A good neighbourhood with easy access to infrastructure such as good roads, schools, hospitals, proper drainage and waste disposal systems attract home owners or tenants who can afford the cost of housing in such places. That is why this research looks at housing from this general perspective. Every individual desires to live in a place where the land is secured and therefore free from threats of evictions; where shelter is safe and access restricted to prevent intrusions or safeguarded against communicable diseases; where household facilities are inclusive within the home to ensure satisfactory usage and privacy and, above all, where community infrastructure is well-provided to easy to access. The chapter therefore discusses housing in general, how residents in the four communities access resources like land, housing, housing facilities and community infrastructure. The sections below discuss the different means by which general housing is accessed in the study community, the non-state actors present, and the rules that regulate access.

To further understand the flow of the discussion, the connection between land, housing, housing facilities and infrastructure in the communities needs some explanation. In many ways, the means of accessing a particular livelihood resource is similar in each of the communities, especially regarding land and housing access. Where land is accessed on reciprocal basis, the same applies to housing and where land access is commodified, housing access operates in a similar manner. There are slight differences regarding access to facilities which are generally commodified in all the communities and have collective usage. Community infrastructure provision procedures also differ slightly in each community.

11.5.1   Local Associations and Land Acquisition in the Study Communities

Land tenure security is important in poor communities. The study reveals that land tenure security differs in the four communities. Whereas residents of Old Fadama squat on a piece of land belonging to the state, the land of Agbogbloshie is in three parts; the market area belongs to the state which rents the space to traders, the main residential part originally belonged to the Gas[2] but much of it has been sold to migrants from other parts of the country, and the land along the railway belongs to the Ghana Railway Company which rents the space to the residents. In Penkye, the

---

[2] Ga is one of the indigenous ethnic groups in Ghana.

land belongs to community members through communal land holding system (this is a system in which land belongs to traditional authorities and families instead of individuals). The land in Akosua Village belongs to the Effutu Traditional Council which gave it to the initial migrants who first arrived in the community. For traders in Agbogbloshie and residents of its main residential areas, as in Penkye, the land is secured and therefore there is no threat of eviction whatsoever. Residents at Old Fadama, Akosua Village and those along the railway line in Agbogbloshie are continuously faced with eviction threats from the land owners; the state, the Effutu Traditional Area and the Ghana Railway Company respectively.

One important feature identified in the study is the market value placed on land in the Accra communities. In Penkye and Agbogbloshie (main residential part) where the land belongs to the residents, the study shows different means of accessing the land. The land in Penkye is regulated through the communal land ownership system. Eligibility to access the land is by affiliation to the *Gyarteh-Otuano* lineage. Part of Agbogbloshie land is largely commodified and, therefore, the communal land holding system is inoperative in that section. A comparison between the two squatter communities again revealed different ways land is allocated. In Old Fadama, there is commodification of land which theoretically is state-owned. Thus, the land is commodified, but the study reveals a different kind of market system at play in the community; the land pricing is much lower partly because of land insecurity and partly because ethnic social relations induce a reduction in pricing of land. In Akosua Village, land allocation among residents is free, similar to communal land holdings regulated on reciprocal basis.

In each community, land is accessed based on different eligibility criteria. In Old Fadama, every migrant can access land if the person has money to pay it. But the amount paid for a piece of land depends on the social relations and ethnic compatibility with migrant land owners. Profit-making is not the sole driver of land sales in Old Fadama. Creating lasting relations for future reciprocities also drives land sales according to the respondents. In Agbogbloshie, Gas can easily access the land because they are Gas. Migrants can buy land from Gas, but others depend on renting land from AMA or the Ghana Railway Company. In each case, rents are paid to the responsible authorities. For those who can access the land because they are indigenes, eligibility is based on linkage to the ethnic Ga leader called *Nii Tackey Commey*, according to community leaders. Most such people have sold their lands to migrants.

According to community leaders in Penkye, eligibility to acquire land there is based on blood linkage to the *Effutu*[3] ethnic group, specifically linkage to the *Gyartey-Otuano* families. There is a communal ideology which prohibits sale of land outside these groups. The ethnic and kinship rules link residents closely to the land as one people. This is strongly backed by cultural values based on 'us instead of them'—the Effutus instead of migrants. The reciprocity system in Akosua Village is

---

[3] An indigenous ethnic group in Ghana.

also based on a communal belief that it is a taboo to sell land belonging to the 'sea god' according to community leaders.

The above discussions points to the fact that land acquisition is greatly influenced by institutional rules. For someone with money, it is relatively easier to access land in some parts of Agbogbloshie and the whole of Old Fadama. However, no amount of money can get an outsider access to land in Penkye or Akosua Village. Family and ethnic affiliations are what are needed there. Also, in Agbogbloshie, market space can easily be rented from the AMA, but the permit to use the space is sometimes dependent on the decision of the trade associations to either allow or prevent the entrance of new traders into the market.

An important feature of poor squatter communities is the constant threat of eviction. Residents in these areas are continuously threatened by eviction. The consequences of forced eviction to those affected have been discussed widely in the literature (Rahman 2001; Du Plessis 2005), yet residents in these communities are regularly threatened with this, which usually will lead to multiple rounds of negotiation between stakeholders. Leaders from the communities employ a variety of strategies to dialogue with the land owners through collective and individual negotiation processes (Patel and Mitlin 2009). There are various organisations involved in the negotiation processes. The residents negotiate with the ruling party to prolong their stay, or else they register their protest against the eviction through an opposition party. And though Old Fadama for instance is portrayed in the literature (Afenah 2012; People's Dialogue 2010) and the media as "the community versus the state," an analysis of the community's existence from the perspective of the people negotiating with the land owners brings a different equation to mind. What if most squatter communities exist because the community leaders "dine" with the state (see also Paller 2015)? In squatter communities, the ultimate decisions remain unknown; residents therefore fear their livelihoods hang in the balance as they await the decisions of their respective landowners. Some argue that it is difficult for evictions to occur in these well-established squatter-communities. This is because largely, political parties recruit their foot soldiers from poor communities. They fear losing votes should they carry out such evictions. Thus, the decision to evict them usually lingers.

11.5.2 Local Associations and Housing in the Communities

Accessing housing is similar to the means of acquiring land in all the communities. Generally, it is expected that access to housing in indigenous communities is based on the rent-free family compound housing system where individuals are allocated free rooms in compound houses. But the research reveals this assertion is only true in the indigenous communities in the country, that family compound housing allocation in Agbogbloshie differs from what prevails in those communities. Squatter settlements are also known to be largely made up of owner-occupied makeshift structures. This assertion is largely true for Akosua Village, but different from what prevails in Old Fadama where some structures are built for sale or for rental purposes (See also Danso-Wiredu 2016, 2018).

The research reveals various types of housing are available in the communities. Some houses are made of makeshift or permanently built single room structures. Others are also permanent or temporary compound houses. The single structures are mostly owner occupied but there are also some for rental purpose mostly as group rooms. Most of the compound housing is occupied by family members of the owners on rent-free basis, but there are also cases where some rooms in the compound houses have been rented to either families or groups of people in groups. The details are explained below.

In Old Fadama, room sizes range between 9 and 18 m^2. They are built mainly of wood and cement blocks, clustered together with no spaces between the structures. They are only use as sleeping places for the residents since there is no space for any other activity.

Houses in Penkye are of the indigenous compound type, where respondents live in rooms located in compounds. The rooms also largely serve solely for sleeping. The houses are family owned; they are occupied on rent-free basis by extended family members of the deceased owners who are sometimes not known to household members. Men and women usually live in separate houses, likewise husbands and wives.

Houses in Akosua Village are rural in nature, mostly made up of earth, bamboo and wood, with a few of the earthen ones plastered with cement. They all have thatch roofing. The houses are mainly detached. They are occupied by owners and rent-free relatives and friends of owners who share household facilities like the compound housing system. Even though these houses are spacious compared to those in the other communities, there is congestion in the rooms, mostly more than four people share a single room in a compound.

Strategies to acquire housing in each of the communities differ. In Old Fadama, most residents own their structures which they have either built themselves or bought, the rests are mostly family and group tenants. Money is needed to acquire housing but also important is social capital to maneuver through the paying processes and bringing together groups of tenants. People who own multiple structures rent some out, they are mostly community leaders who use their leadership power in addition to financial capabilities and social relations to acquire more structures. New migrants and the kayayei[4] are known to rent structures in groups to reduce the amount of rent paid. There are also a few new migrants who mostly use social capital to access temporarily free sleeping places.

In Agbogbloshie, the rooms are much spacious than the ones in Old Fadama. Relatives of the owners of the compound houses live in the houses rent-free. Some rooms in the compound houses are rented out to traders either in groups or families. The makeshift structures belong to migrants who usually build the structures themselves, some are also rented out to family or group migrants. What is common to the

---

[4]Female head-potters.

renting process in both Old Fadama and Agbogbloshie is the flexibility in the processes of renting rooms, which allow weekly payments contrary to the yearly or even more advanced payment systems operating widely across Ghana.

Housing in Penkye is purely accessed on family lineage and rent-free basis, only social affiliation is needed to access land in the community. Rooms in the compound houses are shared among family members. There is gender division as to who lives where; usually, men live separately from their wives.

In Akosua Village, both men and women live on the same compound. Men with more than one wife live with their wives on the same compound. Most structures are built by the owners or passed on to the owners by family members. Community members who do not have their own houses live rent-free with family and kin relations.

11.5.3 Local Governance Structures for Housing Facilities and Utilities in the Communities

Access to household facilities is similar but not the same in the four communities. In all the communities, household facilities are paid on the spot per usage, as are utility bills paid. Utility bills are paid on the spot by multiple residents, rather than being paid each month, which was the norm in Ghana until a few years ago when a prepaid system was introduced in some parts of the country. From the study, it is clear that the accumulated cost of utilities is expensive, but the on-the-spot system makes accessing them somehow affordable and convenient to the users. For example, a few homes in all the communities have meters from the Electricity Company of Ghana (ECG). These homes then share their meters with other homes who help them offset the high cost of electricity bills. This process seems dangerous, yet, it is the most convenient way to deal with the power problem in the communities as also discussed by Silver (2014). All those residents needed is their ability to pay for the facilities. In the Winneba communities, the assemblymen are able to pressure the Effutu Municipal Assembly (EMA) and NGOs into providing these facilities for the residents.

In Old Fadama where the structures are too small for any activity other than sleeping, all household facilities are accessed outside the home where they are used collectively by the community. Toilets, showers, kitchens, water and light are bought from private individuals and an NGO called Ghana Federation of the Urban Poor (GHAFUP). In Agbogbloshie, Penkye and Akosua Village, some facilities and utilities like kitchens, showers and electricity are available at the household level to those who can afford them, but in many cases, they are also sold at the community level by private individuals as in the case of Agbogbloshie, and by the district assembly in the case of Penkye. In all three communities, toilet facilities are collectively accessed at the community level, being sold either by the district assembly (in Penkye) or by private individuals (Agbogbloshie). In Akosua Village, the toilets were built by an NGO so residents were paying no fee for using them at the time of the research. However, due to lack of maintenance by the NGO and community members, the facilities were out of use at the time of the interviews.

The rules for accessibility are based on commodification. Community members able to pay for the facilities and utilities can access them. Also, private individuals or organizations willing to invest in the provision of any facility in Old Fadama and Agbogbloshie are allowed to make the investments, provided the OFADA and the landlord association accept the places they are located. In the Winneba communities, it is the EMA which provides the facilities, but they operate in the same way as what prevails in Accra. There is an agreed amount paid per use in all the communities. The prices are regularly reviewed by the people involve, usually decided by the owners and imposed on the consumers. For these reasons, there are organizations formed by all shower and toilet operators in Old Fadama and Agbogbloshie to ensure uniform charges on such facilities.

11.6 Infrastructure Access and Governance in the Communities

Infrastructure is provided by different actors in the communities. In Old Fadama, because of the recognition of non-state actors leading the community, basic infrastructure like schools is operated by private individuals. Roads and drainage systems are constructed by GHAFUP or private individuals initiated by OFADA. GHAFUP operates the waste collection service in some parts of the community by providing bins to shops and putting dumpsters at vantage points in the community. In the other three communities, it is the responsibility of the municipalities to provide infrastructure but because of financial constraints, these are mostly underdeveloped. NGOs and local associations help at times. Traders for instance, play important roles in providing market spaces at Agbogbloshie market (See also Danso-Wiredu and Sam 2019). Table 11.3 summarizes the various organizations, institutions and individuals involved in the provision of the itemized infrastructure in the study communities.

11.7 Conclusions

This study is about the governance of general housing access in poor communities in Ghana by non-state institutions (summarized in Table 11.4). The study employed the social capital approach and participatory development concept to study the local governance structures in the study communities. The study focused on poor housing in urban communities. To understand the forms of non-state governance structures in the study areas, the chapter gives a general overview of the different associations in the community which are in the forms of established institutions and organizations.

Table 11.3 Redistribution of community infrastructure

	Old Fadama	Agbogbloshie	Penkye	Akosua Village
Drains	OFADA/individuals/trade associations	AMA	EMA	–
Roads	OFADA and GHAFUP	AMA	EMA	EMA
Street lights	–	AMA	EMA	EMA
Waste disposal	Kayabola/GHAFUP	AMA	EMA	–
Schools	Individuals	–	–	EMA
Market structures	–	AMA/Trade Associations	EMA	–

AMA Accra metropolitan assembly, *EMA* Effutu municipal assembly
Source: Danso-Wiredu 2016

Table 11.4 Comparing identified features of the study communities

Comparative features	Old Fadama	Agbogbloshie	Penkye	Akosua Village
Land-tenure security	Squatter-state vested land	Individual ownership; Squatter; state rental	Communal ownership	Squatter
Land-means of access	Commodified	Largely commodified	Free	Free
Land-years of stay	Recent migrants, from 1990s	Native land; over a third-generation migrants	Native land	Over a third generation of migrants
Housing-types	Single structures (more temporary, a few permanent)	Indigenous compounds; Temporary structures (single and double rooms)	Indigenous compounds	Makeshift compounds
Housing-means of access	Group rental; self-built; bought	Rent-free; group rental; self-built; bought	Rent-free	Self-built; rent-free
Population characteristics	Migrants (mostly from the three northern regions)	Native gas; migrants (mostly Akans)	Native Effutus	Migrants (from the Volta Region)
Dominant employment type	Trading; Entrepreneurship; selling of labour	Trading; selling of labour	Fishing (deep sea with canoes and outboard motors)	Drag net on-shore fishing
Local associations	Mimic traditional chieftaincy system; Ethnic associations; Economic associations; Community development association	Economic associations; Welfare associations; Community development association	Traditional chieftaincy system; Traditional norms and beliefs; Welfare associations; Economic associations	Ethnic associations; Economic associations; beliefs in the Ewe tradition and that of the Effutus

The local associations exist and have much influence on land, housing, housing facilities and community infrastructure acquisition processes.

In discussing the strategies employed by the study communities in accessing these facilities, it is clear from the study findings that the non-state actors at the local levels are able to lead the provision of these facilities for the residents to survive. But the reality is that, what exist in the communities are not enough to measure up to the level deemed fit to be counted sustainable. This study therefore acknowledges the need for local people to lead their development including people living in poor communities, but strongly advocate that the state should be part of the main stakeholders to ensure participatory development in these communities, by developing policies that are capable of improving the housing and community infrastructure in poor communities.

References

Afenah A (2012) Engineering a millennium city in Accra, Ghana: the old Fadama intractable issue. In: Urban Forum, vol 23. Springer, pp 527–540

Anyidoho NA (2010) 'Communities of practice': prospects for theory and action in participatory development. Dev Pract 20(3):318–328

Berg BL, Lune H, Lune H (2004) Qualitative research methods for the social sciences. Pearson, Boston

Cassiman A (2008) Home and away: mental geographies of young migrant workers and their belonging to the family house in northern Ghana. Hous Theory Soc 25(1):14–30

Cassiman A (2010) Home call: absence, presence and migration in rural northern Ghana. Afr Identities 8(1):21–40

Chambers R, Conway G (1992) Sustainable rural livelihoods: practical concepts for the 21st century. Institute of Development Studies, Brighton

Cornwall A (2003) Whose voices? Whose choices? Reflections on gender and participatory development. World Dev 31(8):1325–1342

Danso-Wiredu EY (2016) The puzzles of living in urban poor communities: the role of informal governance structures in housing provision in Accra and Winneba. KULeuven, Ghana

Danso-Wiredu EY (2018) Housing strategies in low-income urban communities in Accra, Ghana. GeoJournal 83(4):663–677

Danso-Wiredu EY, Sam EF (2019) Commodity-based trading associations at the Agbogbloshie market in Accra, Ghana. Ghana J Geogr 11(2):1–26

De Boeck F, Cassiman A, Van Wolputte S (2010) Recentering the city: an anthropology of secondary cities in Africa

Du Plessis J (2005) The growing problem of forced evictions and the crucial importance of community-based, locally appropriate alternatives. Environ Urban 17(1):123–134

Effutu Municipal (2019) General profile of the Effutu municipality. Effutu Municipal Assembly, Winneba

Farouk BR, Owusu M (2012) 'If in doubt, count': the role of community-driven enumerations in blocking eviction in old Fadama, Accra. Environ Urban 24:47–57

Ghana Statistical Service (2012) 2010 Population & Housing Census: summary report of final results, Ghana's population Census. Ghana Statistical service, Accra

Ghana Statistical Service (2017) Urbanisation in Ghana. Ghana Statistical Service, Accra

Ghana Statistical Service (2021) Ghana Statistical Service (GSS) released a Preliminary Report which gives the provisional results from the 2021 Population and Housing Census (PHC)

Government of Ghana (2014) Districts in Ghana. Government of Ghana, Accra. http://www. ghanadistricts.com/home/

Helmke G, Levitsky S (2012) Informal institutions and comparative politics: A research agenda. International Handbook on Informal Governance

Hyden G (2008) After the Paris declaration: taking on the issue of power. Dev Policy Rev 26(3): 259–274

Messer N, Townsley P (2003) Local institutions and livelihoods: Guidelines for analysis. Food & Agriculture Organization, Rome

Mohan G, Stokke K (2000) Participatory development and empowerment: the dangers of localism. Third World Q 21(2):247–268

Moreno EL, Head CMB (2011) Living with shelter deprivations: Slums dwellers in the world. Population distribution, urbanization, internal migration and development: An international perspective, 31

Neuwirth R (2007) Squatters and the cities of tomorrow. City 11(1):71–80

Opare JA (2003) Kayayei: the women head porters of southern Ghana. J Soc Dev Afr 18(2)

Paller JW (2015) Informal networks and access to power to obtain housing in urban slums in Ghana. Afr Today 62:30–55

Patel S, Mitlin D (2009) Reinterpreting the rights-based approach: a grassroots perspective on rights and development. In: Hickey S, Mitlin D (eds) Rights based approaches to development: Exploring the potential and pitfalls. Kumarian Press, Sterling, pp 107–126

Patton MQ (2002) Two decades of developments in qualitative inquiry a personal, experiential perspective. Qual Soc Work 1(3):261–283

Peoples Dialogue (2010) Community-led enumeration of old Fadama Community, Accra, Ghana. Housing the Masses, Ghana

Putnam RD (1995) Tuning in, tuning out: the strange disappearance of social capital in America. PS: Polit Sci Polit 28(4):664–683

Rahman MM (2001) Bastee eviction and housing rights: a case of Dhaka, Bangladesh. Habitat Int 25:49–67

Ritchie J, Lewis J, Nicholls CM, Ormston R (eds) (2013) Qualitative research practice: a guide for social science students and researchers. Sage, Thousand Oaks

Serageldin M, Driscoll J, Solloso E (2003) Partnerships and targeted programs to improve the lives of slum dwellers. In: International conference on sustainable urbanization strategies, Weihai

Silver J (2014) Incremental infrastructures: material improvisation and social collaboration across post-colonial Accra. Urban Geogr 35:788–804

Stinchcombe AL, March JG (1965) Social structure and organizations. Adv Strateg Manag 17:229–259

Van der Geest K (2011) The Dagara farmer at home and away: migration, environment and development in Ghana. African Studies Centre, Leiden

Chapter 12
Women and Urban Governance in the Arab World: Case Studies from the MENA Region

Safaa Monqid

12.1 Introduction

Arab societies have undergone political ruptures and social transformations that have resulted in mobilisations identified or associated with emblematic squares or places. Demonstrations in Cairo, Tunis and other Arab cities are spectacular illustrations of this, and the demands are as much about overcoming authoritarian regimes as they are about the equal consideration of women and men in order to establish political systems based on a symmetry of status (Hammouche and Monqid 2022). Women have indeed marked their presence on the political and media scene. They have shown their ability to organise themselves into networks for the respect of their rights and against all forms of violence and social or political discrimination against them, even if in many Arab countries they remain underrepresented in the decision-making process because of certain obstacles, especially cultural ones, which hinder their commitment. The fall of Mubarak, for example, would not have been possible without the uprising of the urban masses who made Tahrir Square the symbol of their movement and without the mobilisation of women who, defying taboos and stereotypes, imposed themselves as essential actors in these demonstrations. Women have been able to strengthen their political participation in local governments from which they were excluded for a long time. They now play a decisive role in the cities of the Arab world through a variety of structures, including NGOs, groups, local authorities, and the grassroots structures of political parties where they express their opinions and specific needs. My presence in the field during the social movements in Egypt in 2011 allowed me to measure the extent of the participation and commitment of women of all ages, all social categories and all tendencies that

S. Monqid (✉)
Université Paris 3, Sorbonne nouvelle, Paris, France
e-mail: safaa.monqid@sorbonne-nouvelle.fr

have asserted the weight of politics in gender inequalities. The same is true of their mobilisation, in 2007, against the insensitivity of the Egyptian power to the suffering of the people during the bread crisis. Crises of subsistence are often triggered and animated by women who occupy the front lines in these struggles (Perrot 1998).[1]

The aim of this contribution is thus to show the position of women, often "invisible" or "unnoticed" actors, in governance. By governance, I am referring to the process that grants a share of power in government policies, from their design to their implementation, to the various actors (Lacroix and St Arnaud 2012). Women have always played an essential role in family, community, local and national organisations. Their collective action has always been important even if it is not always formal. Despite the various obstacles they encounter, we will highlight the changes and innovative strategies that they have adopted in the urban environment in order to transform it. Two main areas will be emphasised: women's empowerment and their access to rights and decisions in public life; and their methods of appropriating public spaces.

12.2 Women's Empowerment: From the Private Sphere to Access to Rights and Decisions in Public Life

Until the end of the nineteenth century, public places in the Arab world were generally reserved for men, and sharp spatial divisions intersected with gendered segregation. Women were assigned to domestic settings for religious, cultural and social reasons. Only necessity and poverty could justify a woman leaving her home. These societies, whose relationships were structured by patriarchy and its conservative values, then suffered disruptions associated with nationalist, anti-colonial and feminist struggles, causing significant changes in the status of women (Lelandais et al. 2019). This is the case of Egypt during the interwar period. Cairo, the first city of the Arab world at the time—modern, cosmopolitan and pioneering—occupied, indeed, an important place in the movement to empower women. Many women took part in the movement, which was backed by nationalist forces. Due to the history, demographics, and cultural and religious significance of Cairo, the city was seen as a model of women's renaissance (Monqid 2016). Initially, the precursors of the Egyptian women's empowerment movement were men of the intellectual and bourgeois elite They were convinced that the status of women had to progress if society was to evolve. For these elite men—the most important of whom were Rifâ'at-Tahtawi, Mohammad Abdu' and Qâsim Amîn—women were essential the construction of the nation.

A number of women became famous in the decolonization movement. They protested during the 1919 revolution, the rebellion led by Sa'd Zaghlûl

---

[1] As Perrot (1998) points out, women played the main role in the popular uprisings of nineteenth-Century France. They were the instigators.

(1858–1927) against the British then occupying the country. This was the first mass participation of women in protest. Among women leaders were Hudâ Sha'râwî (1879–1947) and Saiza Nabarâwî (1897–1985) (Fénoglio 1988). This women's mobilisation has remained in the collective memory, as evidenced by one of Hâfid Ibrâhîm's (1872–1932) poems entitled *mudhâharât nisâ* (Women's Protests). In it, he praises the courage of Egyptian women who broke all taboos, went out into the streets, and joined men in the national struggle (Baron 2005). The mobilisation of women only increase after 1919. This period was characterised by the assertion of women's movements and by the rise of an Egyptian feminist movement. By the end of the 1920s, women had seized the public arena and had revealed themselves by removing the veils from their faces. They came out of their isolation and began their conquest of the city and its spaces. Due to differences of background and social condition, not all women participated in this movement in a unified way. (Lelandais et al. 2019). Women are not a homogeneous group. There are very important differences caused by socio-economic background and the rural-urban dichotomy. There are also socio-cultural inequalities within society generally, which mean that women in certain environments enjoy greater freedom and rights. Especially in cities, women are gradually escaping the roles assigned to them by tradition, and are emerging from "seclusion" to become more actively involved agents in the public sphere. Egypt is not representative of the entire Arab world. The women's liberation movement there is certainly longer-standing than elsewhere, due in part to the influence of the West and the presence of foreign communities. We are therefore able to trace, country by country, category by category, a general evolution in women's agency that started in Egypt at the beginning of the twentieth century and gradually spread throughout the Arab world to all social groups, in accordance with various methods and timeframes (Lavergne 2022).

In Egypt, women were involved in the anti-colonial and nationalist movements from their inception. However, later their contribution was forgotten, as witnessed by the subsequent personal status laws, which acted as institutional mechanism of women's exclusion. These laws have legitimised and formalised gender inequality. As Guionnet and Neveu point out, this is not surprising when we know that history, as it has long been conceived, is the story of the actions of men in the public space. The often androcentric historical narratives produced by men neglect the actions of women (Guionnet and Neveu 2009).

An inclusive holistic historical narrative shows otherwise. Women active in the struggles for independence expressed themselves and made their presence felt through a plethora of civil associations. Several women's organisations have emerged in Egypt: *bint al-ard* society (Daughter of the Earth Association, 1984), *Arab Women's Alliance* (Alliance des Femmes Arabes, 1987) and *The New Woman* (La Nouvelle femme, 1991) (El Sadda and Dayan-Herzbrun 2012). Beyond rallying for changes to exclusionary laws, the aim of these associations was to consolidate the status of women in the public sphere. Women have redirected their activities to the associative field and have turned to civil society as a kind of counter-power in order to achieve their goals and act for change. Egyptian women also distinguished themselves in opposition movements such as *Kifâya* (Enough), where they were actively engaged against the authoritarian regime of Mubarak.

Yet newer groups mobilized in defense of women's rights following the social movements of 2011. These included such associations as *baheya ya masr, nazra, al-mar'a al-jadîda, al-mar'a wa-ddhâkira*, ect. Activists from the various parties involved in the January 25 2011 movement coupled their mobilization in the street with public awareness campaigns for more justice and social equality. Secular and Islamist, young and old, all women have rallied against the lack of public services and the decisions to destroy or expropriate their homes. This is the case with Qursâya, Dahab and Warrâq Islands in the Nile which are coveted of wealthy investors from the Gulf countries (Barthel and Monqid 2011). Women have harnessed social networks and new media to communicate, organise protests and give visibility to their movements. These new technologies have also highlighted their ability to organise themselves into networks and to take militant action. These digital networks offer alternative spaces of freedom of being and expression for women. This is what happened when Asmaa Mahfûz, a 26-year-old blogger and member of the "6 April" youth movement, posted a message urging her male and female compatriots to join the protest in Tahrir Square on 25 January 2011. The same is true of mobilisations against violence against women in the street (Kreil and Monqid 2008).

In terms of achievements, women's struggles have gradually changed the Egyptian legal system. This was the case of the 2000 law giving women the right to *khul'* (divorce), and of the 2004 and 2005 laws granting women custody of children until the age of 15 and equitable and rapid access to alimony. In 2008, the age of marriage was raised from 16 to 18 years and a law penalising female genital mutilation was approved. In addition, the Egyptian constitution of 2014 was presented as an exceptional step forward for women's rights. The state committed, for the first time, to achieve equality of women and men in electoral offices and in the core positions of the civil service (Bernard-Maugiron 2017). A law on sexual harassment was also passed, as well as one granting the right for Egyptian mothers to pass their citizenship on to their children.

Women's access to schooling and employment, rampant urbanisation, public policies in favour of women and improved living standards, particularly in urban areas, have favoured women's access to urban governance. Women have become key players in development and an important force of modernisation in society. They have entered territories and fields that were reserved for men, such as the political field. The members of the *Ba'th* party, In Syria, have made the issue of women's status central to their discussion on the "awakening" of Arab societies. It was in Syria that the first female deputy ministers and attorneys general in the Arab world were appointed (Lavoix 2010).[2] The regime has ensured that women are represented in *Ba'athist*, governmental and parliamentary institutions, even if social recognition

---

[2]Out of the 250 members of the new 2007 legislation, there is at least one member from each electoral district.

of the political role of Syrian women has not yet been achieved. In 1957, Egypt was the first country in the Arab region to elect women to parliament. The 1971 Constitution was amended in 2007 to allow for a quota for women's participation in parliament. For example, under Mubarak, in 2009, the introduction of female quotas would have granted women 64 seats out of 518, or 12% of all deputy positions (Bernard-Maugiron 2017). Under El Sisi's presidency, instructions have been given to promote empowerment and gender equality. As a result, in 2018, out of 33 ministers, eight are women, a first in the Arab world as far as numbers are concerned. We also witnessed the appointing of the first female judge of the Egyptian Criminal Court and of a female president of the Economic Court, among others. In 2017 and 2018, the first female governors were appointed in the provinces of *El Beheira* and *Damietta*, as well as a female vice governor of the Central Bank.

Similarly, in Jordan, a quota system was introduced in 2003, reserving 6 of the 110 seats in parliament (or 5.4%) for women. In 2010 this quota was increased from 6 to 12, or 10.9%. In the elections of November 9, 2010, out of a total of 120 seats, 13 Jordanian women were elected, 12 of them through a quota. They now hold 10.8% of the seats in parliament. In July 2011, a new law on municipalities increased the quota for women on municipal councils from 20 to 25 per cent. There was a slight increase in the representation of Jordanian women on local councils from 1.6% in 2002 to 4% in 2008. Women are also present in trade unions, as is the case in the textile and garment sector in Jordan, where women reportedly represent about 60% of union members. In the 2010 executive board elections, they managed to occupy 5 out of 9 positions (Lelandais et al. 2019).

Such measures would not have been possible without the efforts of women and their involvement in public affairs. They are also beginning to represent a trend in the Arab world with the revision of family laws.[3] Due in part to urban development and the widespread use of new technologies, the condition of women in the Arab world as a whole is tending to become more homogenous as a result of the standardisation of living conditions and lifestyles and in accordance with models that are spreading across the world (Lavergne 2022).

---

[3]In the Maghreb, the process of changing women's rights in the family has been more or less complicated, depending on the history and political system of each country. The battles of Moroccan women over several decades finally led, in 2004, to the modification of the Family Code (*Moudawana*) to remove the wives' duty of obedience and introduce joint family responsibility and legal marriage age. The Algerian Family Code was finally changed in 2005, but it still considers women as minors (Lelandais et al. 2019).

12.3   Ongoing Mobilisation in the Struggle Against Gender Discrimination

In spite of significant social changes, gender inequalities continue in many Mashreq countries and women continue to suffer significant discrimination due to substantial social and cultural obstacles. Women are generally marginalised, excluded from important political positions and decisions, and even when they are elected, they find themselves in the midst of government ministries directly affecting women. Women also remain poorly represented in trade unions. In Jordan, they constitute only 20% of the trade union community, which remains a male prerogative, especially since it is not widespread in all sectors of activity. The place they occupy is ridiculously small (at the very bottom of the ladder and excluded from important decision-making positions), except in the textile sector where they are very present and active (Lelandais et al. 2019).

In the private business sector, gender inequalities in terms of status, salary and access to training and promotion remain entrenched. Women are also affected by the lack of employment and job security. The most widespread form of female job insecurity is informal work, which is increasing due to unemployment. We can still recall the joint actions of female workers in the public textile factories of Mahalla, a large single-industry city in the Egyptian Delta and a major center of workers' unionism. The initiators of the great strike of 2006–2007, Wedâd ed-Demerdach and Amal Saîd, denounced the discrimination against women (including large wage gaps compared to men, insecure employment, harassment and threats of rape. . .). This event lies at the origin of the 2011 movement.

In addition, in many Mashreq cities, women have to face various types of violence against them such as sexual violence and institutional violence, including police intimidation and targetted physical attacks during demonstrations. The purpose is to exclude women from public and political spaces. They are victims of discriminatory laws (Egypt's various constitutions, with few exceptions, do not guarantee the principle of equality between women and men and impose male guardianship). Women are also victims of family violence, such as early and forced marriage, female genital mutilation, and so-called "honour killings." Murderers who claim that they are defending the family's honour are given reduced sentences. Women and children are also the primary victims of wars in the region because of all the economic, social, cultural, security and health problems wars generate. This is the case of Palestinian, Syrian, Iraqi and Yemeni women who have seen their rights decrease due to conflicts (rising illiteracy figures, malnutrition and no access to health care, increasing number of rapes, kidnapping and trafficking of women, murders in the name of religious morals. . .).[4] Women also face religious extremism that instrumentalizes the cause of women by denying some of their rights and sending them back to their traditional function (Lelandais et al. 2019).

---

[4] Several women working in the public sector, doctors, vets, civil servants... were shot or decapitated.

12.4   Women and Ways of Taking Over Urban Spaces

Monitoring urban situations in the cities of the Mashreq as well as in the Maghreb provides information on gender-differentiated practices, which E. T. Hall has called the hidden facets of space (Hall 1971). These hidden facets reveal invisible discrimination and relations of economic, political, social and symbolic domination, including gender-based discrimination (Denèfle 2004).[5] Each gender behaves differently in public, they do not go to the same places at the same time, they do not take ownership of the neighbourhood or the city in the same way. Some places dictate behavioural norms that generate gender inequalities. There are several territories in the city, male and female territories, territories that are allowed, tolerated, and others that are forbidden. . . Women have less control over the city even though they have gained massive access to public space. The space outside the home remains occupied by the traditional private/public dichotomy. Women encounter obstacles and very clear resistance when they enter the public sphere (Monqid 2014). One of the most socially unacceptable of obstacles is certainly the unsafe and violent experiences they may encounter in public places, including hostile comments, inappropriate gestures and verbal and/or physical assaults, which breach their intimate, mental and physical space. Such is the case of Taḥrir Square, which has become the symbol of the gender violence that afflicts Egypt. The systematic gang rapes that took place in this square shed international light on the daily sexual harassment experienced by many Egyptian women, veiled or not, especially in public transport.[6] In order to avoid such attacks, there are women-only carriages in the Cairo metro for those who do not wish to mix with men. And although a law against sexual harassment on the street was enacted in 2014 and constitutes progress, it is doubtful that it will be enough to change the mindset overnight (Fortier and Monqid 2017).

These assaults on women raise concerns about the female presence in public space as well as the issue of insecurity experienced by women and overall urban safety. It also brings the discussion of the legitimacy and/or illegitimacy of women's presence in the public space into focus. These realities obviously influence their means of accessing the city and hinder their mobility and their involvement in public life. As demonstrated by their use of space for purely useful purposes, their leisure activities, which are indoor or family activities, and the strategies they use to avoid male spaces and to censor themselves or to conceal themselves and become invisible, particularly through veiling. Generally speaking, in the Arab-Muslim context and beyond, the main obstacle to gender equality remains the issue of control over women's bodies, their sexuality and their freedom of movement. Daily limitations in the name of collective morality are major impediments to women's freedom and mobility. However, even for Western women, access to public space is a recently

---

[5] This concept refers to social forms of organising differences. It is a social construct.

[6] This is reflected in Mohammed Diab's 2010 Egyptian film, titled *678*, which refers to the number of the bus.

acquired right. It is still more or less under male control, if only because of the level of urban violence which affects women far more than men (Hanmer 1977).

In addition, public urban policies are also marked by gendered norms which prevent women from being full citizens and actors in the city. Cities were and still are often designed, organised, managed and produced within a framework dominated by men (architects, city planners, developers, politicians, etc.), since they are the ones who, for the most part, think, produce and impose codes relating to space and the urban environment. Women are mostly excluded from management, development and planning of the city and its spaces (Denèfle 2009). The traditional layouts of the region's cities, the "medinas," clearly reflected the different roles assigned to men and women. Traditional urban space was in fact divided into two strongly differentiated social worlds: an external, public, masculine world and an internal, private, feminine world. The exclusion of women from public life is the most apparent aspect of the unequal relationship between the sexes in traditional society. Women lived in anonymity; they were identified with the home (*dâr*), sacred and inviolable, the domain of intimacy. The small alleys, (*driba*) were the extension of the private space, they protected the most sacred places: the houses where the women lived. The latter were not only protected in the space of the medina with its narrow, dark alleys, but also in the space of the dwelling, which was organised around the principle of closure (Monqid 2014). As Sylvette Denèfle has shown, this masculine worldview is certainly expressed in the general morphology of space. But it also undergirds the planning of territory and urban sites and even in the dissemination of urban culture. This gendered worldview is shared by almost all actors—including women—and functions as a matter of course. As political and social actors, women have therefore not yet established a distinct conception of "living together,"of the inhabited, the habitable and the uninhabitable in social space and time, and they cannot do so without political parity in the creative and decision-making institutions (Denèfle 2009).

Consequently, we are witnessing resistance from cities and urban environments towards the struggle of women who constantly have to fight for their rights, one of the most basic of which is access to public space. The power inequalities between men and women in the private/public spheres are numerous and we cannot talk about good leadership, i.e., democratic management of public affairs, respect for human rights, social justice and decreasing inequalities, if we ignore the needs, interests and difficulties that women have to cope with in the city. It is essential that women have a voice in public debates. Institutional awareness of the need to include women in decision-making bodies must be made effective by adopting concrete measures to combat gender-based discrimination at all levels: political representation; administrative apparatus, justice system and civil society (Denèfle 2009).

References

Baron B (2005) Egypt as women. Nationalism, gender and politics. American University of Cairo Press, Cairo

Barthel P-A, Monqid S (2011) Le Caire. Réinventer la ville. Collection Villes en mouvement. Autrement, Paris

Bernard-Maugiron N (2017) Le statut juridique des femmes dans l'Égypte post-révolutionnaire. In: Fortier C, Monqid S (eds) Corps des femmes et espaces genrés arabo-musulmans. Karthala, Paris

Denèfle S (ed) (2004) Femmes et villes. Presses universitaires François Rabelais, Maison des Sciences de l'Homme. Villes et Territoires, Tours

Denèfle S (ed) (2009) Utopies féministes et expérimentations urbaines. Presses universitaires de Rennes, coll. Géographie sociale, Rennes

El Sadda H, Dayan-Herzbrun S (2012) Droits des femmes en Égypte, l'ombre de la première Dame. Tumultes 1(38–39):299–311

Fénoglio A (1988) Défense et illustration de l'Égyptienne, aux débuts d'une expression féminine, Le Caire, CEDEJ

Fortier C, Monqid S (eds) (2017) Corps des femmes et espaces genrés arabo-musulmans. Karthala, Paris

Guionnet C, Neveu E (2009) Féminins/Masculins, Sociologie du genre. Armand Colin, Paris

Hall TE (1971) La Dimension caché. Seuil, Paris

Hammouche A, Monqid S (eds) (2022) Espaces et genre dans le monde arabe: Des transformations urbaines aux mutations des rapports de sexe. Karthala, Paris

Hanmer J (1977) Violence et contrôle social des femmes. Questions Féministes 1(Novembre 1977): 68–88. http://www.jstor.org/stable/40619104?origin=JSTOR-pdf

Kreil A, Monqid S (2008) Femmes et harcèlement sexuel en Égypte. In: Farag I (ed) Chroniques égyptiennes. Cedej, Cairo, pp 151–166

Lacroix I, St Arnaud P-O (2012) La gouvernance: tenter une définition. Cahiers de recherche en politique appliquée 4(3):19–37

Lavergne M (2022) Femmes dans les espaces transformés du monde arabe: Etat d'une mutation complexe en cours. In: Hammouche A, Monqid S (eds) Espaces et genre dans le monde arabe. Des transformations urbaines aux mutations des rapports de sexe. Karthala, Paris, pp 17–30

Lavoix V (2010) Femmes, pouvoir et voile en Syrie. Hérodote 1(136):100–120

Lelandais G, Monqid S, Semmoud N (2019) Femmes, droits et participation citoyenne au Maghreb-Mashreq. In: Levy C, Martinez A (eds) Genre, féminismes et développement: Une trilogie en construction. Presses de l'Université d'Ottawa, Ottawa

Monqid S (2014) Femmes et villes: Rabat, de la tradition à la modernité urbaine. Presses Universitaires de Rennes, Rennes

Monqid S (2016) Mouvements féminins et féministes en Égypte: Rétrospective et histoire d'une évolution. In: Benzenine B, Monqid S (eds) Femmes dans les pays arabes: Changements sociaux et politiques, vol 74. Centre de Recherche en Anthropologie Sociale et Culturelle (CRASC), Oran., Revue Insaniyat, pp 49–73

Perrot M (1998) Les femmes ou les silences de l'histoire. Flammarion, Paris

Part IV
Resource Management

Chapter 13
Governance of Water Scarcity in the Syrian Jazeera, from Mismanagement to Political Settlements: Case Study of Ras Al Ayn (1950–2020)

Ahmed Haj Asaad and Khadija Darmame

13.1 Introduction

Water is inextricably tied to social and economic stability since it is a critical component for survival and development in arid and semi-arid regions. It is, therefore, necessary to attach importance to studying the rational of water security which is defined by UN-Water (2013) as "The capacity of a population to safeguard sustainable access to adequate quantities of acceptable quality water for sustaining livelihoods, human well-being, and socio-economic development, for ensuring protection against water-borne pollution and water-related disasters, and for preserving ecosystems in a climate of peace and political stability."

Risks associated with natural environment have an impact on water security in Syria's Jazeera region.[1] The area is characterized by an arid and semi-arid climate, with seasonal and annual variations in rainfall, and the average annual rainfall is less than 200 mm over large areas. The region is subject to variations in rainfall within a single season and to periodic years of drought, which can range from 2 to 5 years. With such uncertainty in the rainfall calendar, the fluctuation of agricultural

---

[1] Known as Northeastern Syria in the literature of humanitarian organizations, journalism, and research fields. The Syrian area covers the left bank of the Euphrates River, with the entire Al Hassakah Governorate, and the parts located on the left bank of the Euphrates River in Dayr az Zawr, ar Raqqa, and Aleppo Governorates.

A. Haj Asaad (✉)
Geo Expertise, Geneve, Switzerland
e-mail: ahmed.haj.asaad@geoexpertise.org

K. Darmame
Al Akhawayn University in Ifrane, Ifrane, Morocco
e-mail: K.Darmame@aui.ma

© The Author(s), under exclusive license to Springer Nature Switzerland AG 2024
K. Darmame, E. Ross (eds.), *Local Governance and Development in Africa and the Middle East*, Local and Urban Governance,
https://doi.org/10.1007/978-3-031-60657-1_13

production increases, threatening financial stability of farmers, and any other economic activity that is linked to the agricultural sector.

Moreover, the irrational governance of water resources, which has encouraged intensive use of water for agriculture, has led to water surface pollution, a decline in ground water levels, and an increase in its salinity, causing the drying up of many springs and rivers (Soumi and Danial 2010). Therefore, providing the population with access to potable water has become a serious dilemma, forcing governmental entities to channel water from areas of water availability to areas of needs in cities and rural areas, where local water sources have been depleted (Haj Asaad and Chamali 2016). In many cases, water is channelled over distances exceeding 70 km. As these water networks cross regions controlled by various military and political groups, they are subject to encroachment and used as weapons or threats for political gains.

The focus in this article will be on the post-2011 era, though giving special consideration to the changes that occurred in the past century. The aim is to provide a deep understanding of the historical context of water security rational in the Syrian Jazeera, and its impact on political stability, social security, and economic prosperity in the region. We will analyze the natural conditions that characterize the region, particularly the Ras Al Ayn springs, as well as the economic, social, and political dimensions of investments there. Furthermore, we will examine the population's components and the changes that have been occurring since the beginning of the last century. We will explore the changes in military control that the Syrian Jazeera (northeastern Syria) has experienced, to go then into detail about water governance issues. A specific emphasis will be on the Ras Al Ayn springs. We will review the changes they experienced due to mismanagement, drought, the military-political use made of water supply and access for political purposes, and the impact they had on the social structure in the region.

13.2   Research Methodology

For the research method, we used the analysis of the crisis aspects, water usage, land use development, demographic change, conflict effects, and the role of water in the reconstruction process as a common denominator. The are studied covered the left bank of the Euphrates River in the Syrian territory. It consists of the entire Al Hassakah governorate, and the parts located on the left bank of the Euphrates River in Dayr az Zawr, Ar Raqqah, and Aleppo Governorates. This area is also called Syrian Jazeera, which is currently known as Northeastern Syria in the literature of humanitarian organizations, the press, and research. In this research, we will often use the term "Jazeera," but we may also use the term "northeastern Syria." We will be focusing on Ras Al Ayn springs as a case study, and areas that benefit from it.

The Syrian Jazeera region is located between latitudes 34.34–37.13° North, and longitudes 93.37–42.42° East. The total area that will be studied is approximately

51,100 km^2, of which the cultivated areas account for 30% of the total cultivated area in Syria. It is considered a major source of wheat and cotton in the country. The region includes the Tigris and Euphrates rivers, in addition to many of their tributaries such as Al Khabur, Al-Sajur, Al Balikh, Zarqan, and Jaghjagh. Data for this research was collected from many source: official documents (Syrian government, international organizations, and civil society organizations) including: maps (topographic, geological, and soil), development plans, reports, as well as from articles and books. Some of the data was collected from media and social networks and has been verified and validated by our experts in the field. It has then been digitized, formatted, and geographically located. In addition to the available data, we carried out 35 interviews using a semi-structured questionnaire with open-ended questions. These interviews took place in September and October of 2020. People interviewed were Syrian experts (engineers, hydrogeologists, geographers, agronomists, and political scientists), government employees, former civil servants, employees of Non-Governmental Organizations (NGOs), international organizations operating in the targeted area, and residents, particularly farmers who use irrigation systems or wells.

These surveys initially focused on the technical, social, and economic characteristics of the Ras Al Ayn area, including the Uluk pumping station. They then addressed the political dimensions associated with local, regional, and international actors, as well as their geopolitical goals in the region. The Geographical Information System (GIS) has also been used as a tool for understanding and facilitating the analysis of the social and economic interactions with water, its geopolitical dimensions, and the political economy at the local and regional levels. We also used GIS to create illustrative thematic maps for conflict zones and their shifts, water facilities, and the region's ethnic and religious distribution.

13.3 The Geographical Situation of Northeastern Syria

Northeastern Syria has enormous natural potential. It is rich in water, fertile land, pastures, oil, and gas. It is also characterized by a diverse climate that enable the integration of agricultural production in terms of crops, vegetables, and animal products. This research will focus on water resources as one of the geographical potentials, particularly Ras Al Ayn's springs and the changes in water use for irrigation and drinking purposes, as well as the effects on the hydrogeological system, which is considered as the backbone of the region's life and economic development.

The Syrian Jazeera is divided into two distinct regions. The lower Jazeera is dry, while the upper Jazeera in the north has a relatively humid climate receiving more than 250 mm of precipitation per year (Fig. 13.1 and Table 13.1). Annual rainfall decreases from the northeast to the southeast; averages reach to more than 500 mm in the Tigris region in the northeast to less than 200 mm to the southeast. This indicates a transition from the semi-arid system to the desert system (the lower

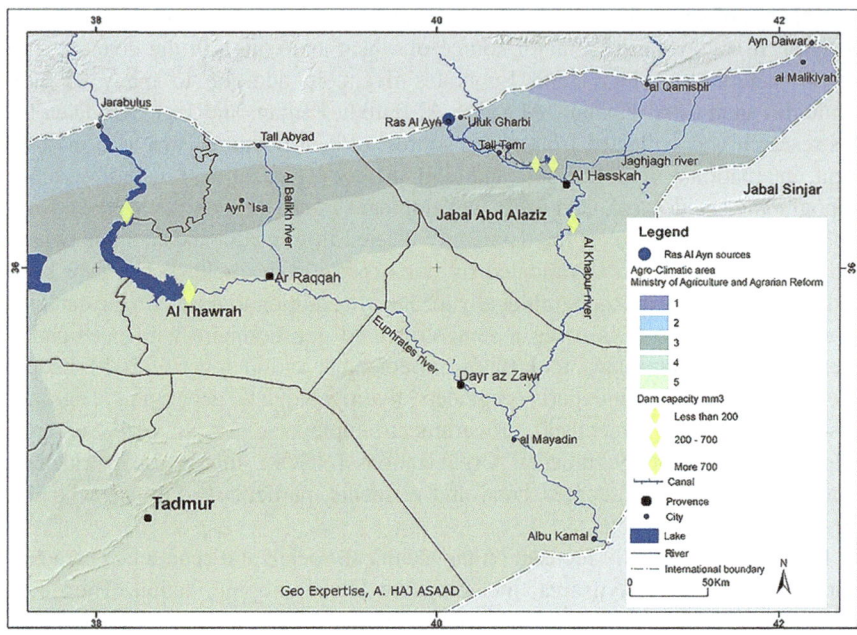

Fig. 13.1 Map of rainfall areas of northeastern Syria

Table 13.1 Legend of rainfall areas

Rainfall Area	Average annual precipitation
Region 1	More than 350 mm
Region 2	250–350 mm (when the rate is more than 250 mm for 2 years within 3 years period)
Region 3	250–350 mm (when the rate is more than 250 mm for 1 year within 2 years period)
Region 4	200–250 mm
Region 5	Less than 200 mm

Source: Republic of Syria, Central Bureau of Statistics 2010

Jazeera) starting from the south, particularly from Abd Alaziz in Syria and the Sinjar Mountains in Iraq.

Al Hassakah lowland is located between the mountains of Abd Alaziz and Sinjar, where the Khabur River, originating at the Ghazal Spring within Turkish territory, flows into Syria to meet Ras Al Ayn springs before reaching the Euphrates, where many tributaries join it. The most important ones are Jaghjag which runs through Al Hassakah city, and Zarqan which runs through Tall Tamr (Gibert and Févret 1953). The Khabur River runs through approximately 320 km of Syrian territory. Ras Al

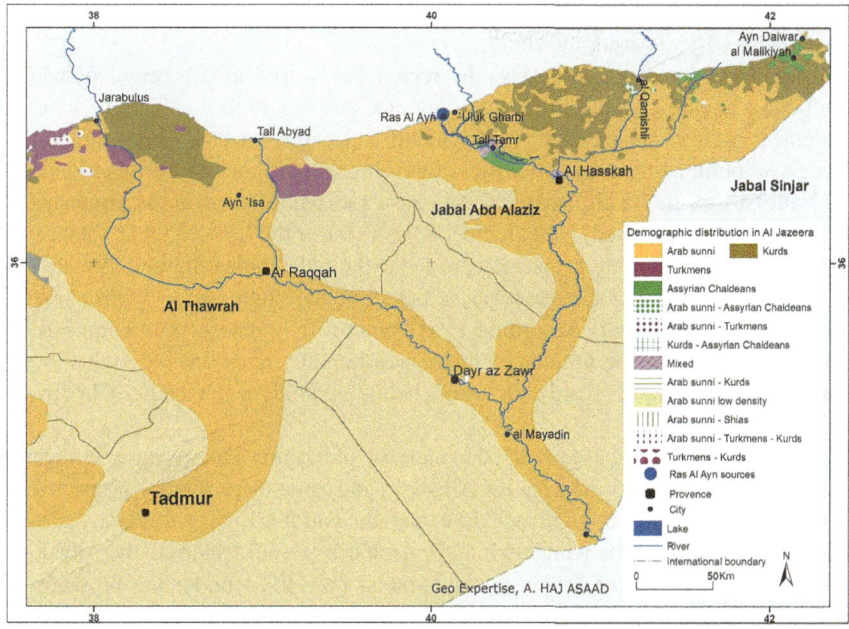

Fig. 13.2 Distribution of demographic groups in Northeastern Syria

Ayn springs[2] are located within the administrative borders of Al Hassakah Governorate, near the Syrian-Turkish border (Fig. 13.1). They are considered one of the most important springs in the Cretaceous-era formation famous for a flow rate reaching 45m^3/second and their regular and continuous flow throughout the year. Along with the Khabur River, these springs represented the driving force not only of Ras Al Ayn, but also for a large part of Al Hassakah Governorate.

13.4   Demographics of Northeastern Syria

Syria's Upper Jazira region has one of the most diversified populations. Arabs, Kurds, Turkmen, Syriacs, Circassians, Assyrians, Chaldeans, Armenians, Circassians, and Mardili, as well as Yazidis, Christians, and Muslims, formed the *Jazrawi* population component (Fig. 13.2), with various ethnic, confessional, and tribal affiliations. Such diversity is due to the strategic geographical location of the region, which is characterized by its fertile soil and abundant water, despite being

---

[2]Among these springs are Ayn Al-Zarqa, Ayn Al-Banos, Ayn Al Hussan, Ayn Doulab, Ayn Al Kebrit (Sulfur), Ayn Salouba.

surrounded by a harsh mountains climate to the north and the arid climate of the desert lands (*Albadeya*) to the south.

Throughout history to this day, the region has served as a meeting point for conflicting empires, feudal lords, and emirates, and has been a subject of various settlement policies as well. In the twentieth and early twenty-first centuries, the successive political-military authorities of the Ottoman Empire, the French Mandate, the Nationalists, the Baath Party, and the new actors whose mission has grown since 2011,[3] have implemented a teleological adjustment in the demographic balance to consolidate their authority in an area opposed to their ideologies and/or existence. As a result of these settlement practices, a demographic composition with a fragile social contract prevailed. According to references, the demographic composition during the period of the Ottoman Empire was largely composed by Arab tribes,[4] nomads, semi-nomadic Kurds, or those living in villages,[5] Yazidis, Christians, and Jews.

Arab and Kurdish tribes alternated in their use of pastures and springs. During the summer, Arabs would come from the Albadeya and camp there, while Kurdish tribes would live in the adjacent highlands. In winter, the Kurdish tribes would move down from the mountains in the winter, while the Arab tribes would return to the Albadeya (Boris 2010). As for the few resident farmers, they lived under the pressure of *al-khawah* (a fee in exchange for protection) and invasions coming from the village and the surrounding mountains. However, in the nineteenth century, the Ottomans started implementing a strategy of protecting the agricultural areas from the nomadic and semi-nomadic tribes' invasions, by settling the Circassians in 1876 (Gibert and Févret 1953).

During the French mandate, the French High Commissioner resorted to a settlement-based population policy. The goal was to shift the demographic balance within the region to ease pressure by anti-mandate opponents, and to stabilize borders with Türkiye (Gorgas 2009). This policy entailed settling Kurdish refugees fleeing the Kemalists (Olson 1989) and Christians from neighbouring countries within Al Jazeera: the Assyrians from Iraq and the Armenians and Syriacs from Türkiye (Gorgas 2009). In this regard, the High Commissioner continued to conduct the vigorous settlement policy for nomads that the Ottoman authorities had undertaken prior to its collapse (Gorgas 2009).

After a decade, these actions have had a considerable impact on Upper Jazeera's demographics. Before 1927, the number of Kurdish villages was barely 45. In 1941, the population of the Upper Jazeera reached 141,390, with of 57,999 thousand Kurds (semi-nomadic and sedentary), 48,749 Arab Bedouins and residents, and 34,945 Christians of various sects and languages (Gorgas 2009).

The withdrawal of the mandate authority gave rise to the Arab movement influenced by the prevailing nationalist theories of that period. Jazeera's

---

[3] The Islamic State of Iraq and Syria and the Syrian Democratic Forces.

[4] like Shammar, Wati, Al-Baqqara Al-Jabbour, Al-Sharabin, and Qais.

[5] Al-Milli, Dakuriyya, Hurakan, Shatayet Al-Hassan and Al-Mira Al-Kurdi

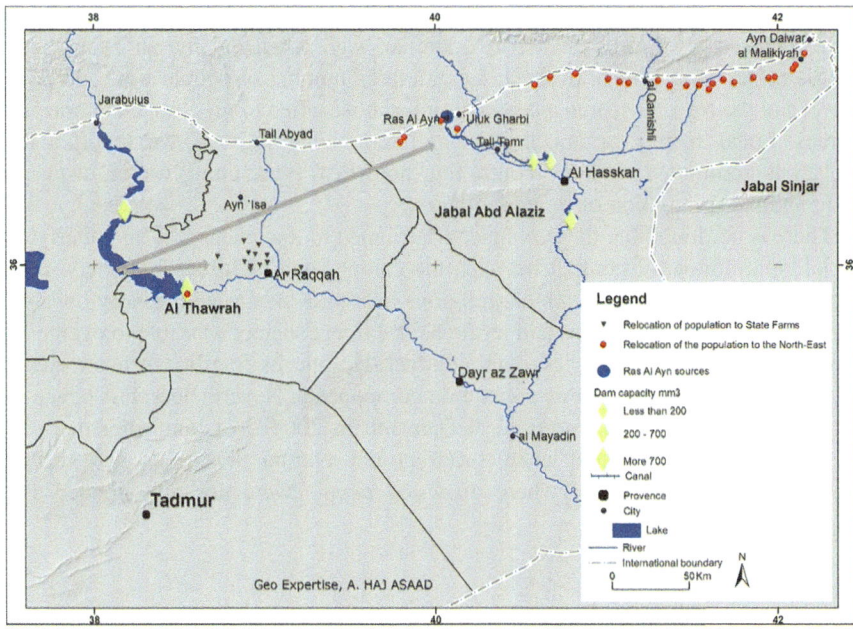

Fig. 13.3 Map of residents' relocation from villages flooded by the Euphrates Reservoir

demographics were restructured at this time, and this is represented in the 1962 census. According to this census, the Kurds in Al Hassakah were divided into "national Kurds" who were registered in the records, and "foreign Kurds" who were not (Al Jazeera News 2009). The 1962 census constituted an exclusionary legal means to deprive Kurdish farmers of benefiting from the Agrarian Reform Law as they were categorized as non-Syrians. The operation to change the population composition continued under the rule of President Hafez al-Assad. Construction of the Euphrates Dam created a 665km^2 reservoir. This submerged 190 villages with their farmlands along the right bank of the Euphrates River that were owned by the Walda tribe. In 1975, after their villages were flooded, around 60,000 Arabs[6] were resettled in the Upper Jazeera, in a majority Kurdish area along the Syrian-Turkish border (Ababsa 2009) (Fig. 13.3).

With the outbreak of the revolution, and after it turned into a military conflict, the Jazeera region evolved into a refuge welcoming the displaced as well as into a source of displacement itself, depending on which military force controls the region. When

---

[6]During their meeting with President Hafez al-Assad in 1974, the sheikhs and members of the Walda tribe first refused to relocate to al Qamishli. After a private and exclusive meeting with President Hafez al-Assad, Sheikh Shawakh al-Bursan withdrew his rejection and provided his approval to move to al Qamishli. Then President Assad asked Sheikh Shawakh along with the sheikhs of Syria's tribes to greet King Faisal Al Saud, King of the Kingdom of Saudi Arabia, during his visit to Syria in 1975.

ISIS took control of parts of Jazeera, there was a large displacement of residents who were replaced by ISIS loyalists. In a similar way, when the Syrian Democratic Forces (SDF) took control of the Jazeera, a large number of people were forced to leave, and the region became a destination for those allied to the Syrian Democratic Forces. Local society organizations and international NGOs have reported the violations committed by those controlling northeast Syria, citing forced displacement and the destruction of several villages.

There is no doubt that these policies contributed to the creation of an underlying social environment of conflict between the components of Jazeera's society, conflicts in which ideological-national-religious hostilities merge with rivalry over the region's wealth, its oil, water, and fertile land. Jazeera society was able to overcome this environment by forging a social contract. Despite its fragility, it has fostered coexistence among the various population components. Nevertheless, this began to deteriorate following the events in al Qamishli in 2004. Not long after that, the Jazeera became a battlefield where sectarian and religious hostilities were incited among the population, and where displaced people were not only robbed and impoverished but killed.

13.5   The Succession of Military Forces and Their Entering Into Alliance in Northeastern Syria Since 2011

The Jazeera region, like other areas in Syria, has had a succession of powers controlling the territory. At the beginning of the popular movement, in 2011, regime forces withdrew from the Kurdish regions and were replaced by the People's Protection Forces (YPG). This was the military arm of the Democratic Union Party (PYD), dominated by the Kurdistan Workers' Party (Balanche 2018). For the Turkish authorities, the PYD is a branch of the Turkish Kurdistan Workers' Party (PKK), which is regarded as a terrorist organisation by Türkiye and many other countries, including the United States and some European Union members. For the other areas of al Jazeera, including Ras Al Ayn, the fell to the Free Army took as the regime and the pro-regime forces fled in November 2012. But it was soon sidelined by Jabhat al-Nusra (now Jabhat Tahrir al-Sham) which seized the area and held it until July of 2013 (Benhaim and Quesnay 2016). The YPG then expelled the Jabhat al-Nusra and took control of Ras Al Ayn. These movements corresponded with ISIS's control of vast parts of Syria and Iraq, including parts of the Jazeera.

At the end of 2014, the Western coalition led by the United States and the YPG concluded a military agreement to fight ISIS. Following this alliance, the SDF were established, including the People's Protection Units and local military factions. Made up of Arabs, Assyrians, and Turkmen, the People's Protection Units remained the main force within the alliance despite Türkiye's opposition to its inclusion, considered they would allow the Turkish Kurdistan Workers' Party to operate along its southern border.

Table 13.2 Distribution of military forces in areas benefiting from Ras Al Ayn Springs and the civil administrations related to water resources management

Military forces	
Local forces	Supporting International Powers
SDF	The United States, France, Britain
National Army	Türkiye
Regime forces	Russia, Iran

The SDF and the Western coalition launched combined military operations against ISIS on January 26, 2015, and declared victory in March 2019. Thus, the SDF took control of the whole left bank of the Euphrates River in Syria. It is the richest agricultural region in Syria, accounting for 30% of cultivated land, and is abundant in water resources, including hydroelectric facilities, and in oil resources as well as cement production.

The advance of the SDF and the disregard of the Western coalition for Türkiye's concerns about the SDF deployment on its borders were critical to the SDF's response. As a result, they collaborated with the National Army to took control of three areas ("safe zones"), with the cooperation of the Turkish army: the Euphrates shield (Dir' Al Furāt region: Jarabulus - Azaz), the Ghosn al-Zaitūn region (Afrin), and Nabaa as Salam region (Tal Abyad - Ras Al Ayn).

Thus, the Syrian Jazeera came under the control of three competing forces (Table 13.2). The National Army with support from Turkish forces took control of 7.9% of the Jazeera's entire territory. The SDF control 91.9%, while the regime forces and their allies were limited to the Security Zones in Al Hasaka and Qamishli, as well as the airport (Fig. 13.4). There are also both international and regional actors whose presence is characterized by the establishment of military bases and deployment of soldiers across the region, as in the case of four of the permanent members of the Security Council: Russia, the USA, France, and Britain. In addition to Türkiye, Iran and the Arab Gulf states offer military, technical and logistical support to groups in the region.

13.6   Water Investment Governance in the Syrian Jazeera: Ras Al Ayn-Uluk Pumping Station

13.6.1   From Abundance to Drought

Al-Hassakah and the villages directly adjacent to the bed of the Al Khabur River depended on its water to secure their daily needs for domestic use and for irrigation as well. In the middle of the last century, the Upper Jazeera's water resources were subjected to a significant exploitation by Syrian and Turkish farmers. The water projects implementation began in Syria in the early 1950s. The Tal Maghas Dam was built on the Al Khabur River, to irrigate 8400 hectares. This dam caused the resettlement of 2100 families, representing approximately 15,000 people (De Vaumas 1952). In the 1980s, the Al Hassakah project was completed on the al

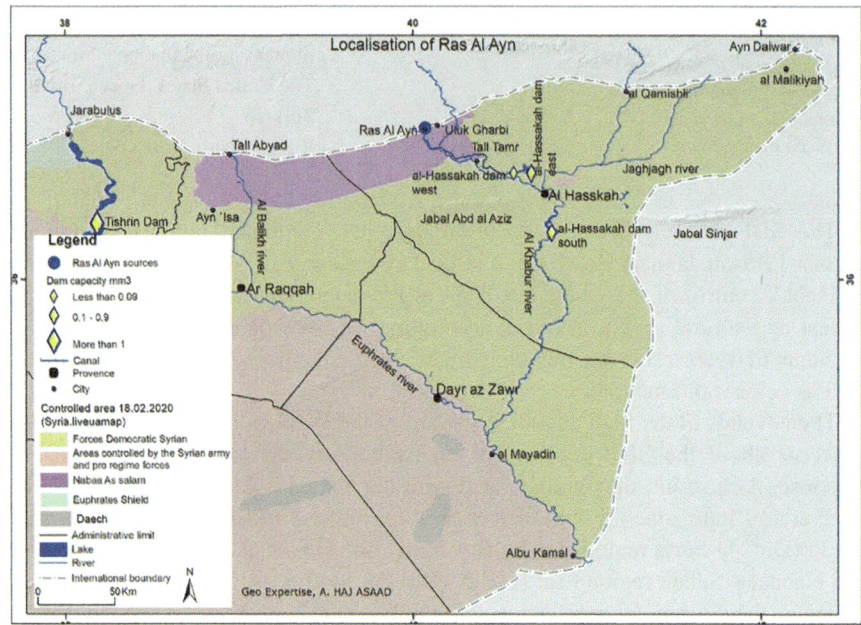

Fig. 13.4 Location of Ras Al Ayn Springs and areas of military control in Feb. 2020

Khabur River, which included the construction of three medium-capacity dams (Fig. 13.5). The storage capacity of the Al Hassakah Dam-West is 91 m³, that of Al Hassakah Dam-East is 232 m³, and the storage capacity of Al Hassakah Dam-South, in the Central Valley, is 665 m³. These structures aimed to irrigate 150,000 hectares with a 200% intensification rate, and to provide water for domestic use to the cities AL Hassakah and Tall Tamr and their rural surroundings. They were supplied with water from Al Hassakah Dam-East to which water was drawn from Ras Al Ayn springs through an open canal 67 km long.

At the national level, the lack of planning and coordination with Türkiye contributed to this project's failure. Prior to the conflict, the region saw an increase in the number of private artesian wells dug for irrigation within the basin direct influenced by the springs. According to Soumi and Danial (2010), the irrigated land area expanded from 2400 to 44,550 hectares while the number of wells on the Syrian side increased from 232 to 2391 between 1984 and 2006. There is a similar increase in well-digging on the Turkish side, where croplands are irrigated with groundwater extracted from areas that directly impact the springs.

The huge extraction of groundwater for irrigation occurring on both sides of the border has negatively impacted the basin and the springs, resulting in the ceasing of free flow of springs in early 2001; their previous flow had approximately 45 M³/s (Fig. 13.6). This caused an increase of water pollution, excessive salt levels, and the growth of toxic plants (lichens) in the reservoir of Al Hassakah Dam-East, giving the

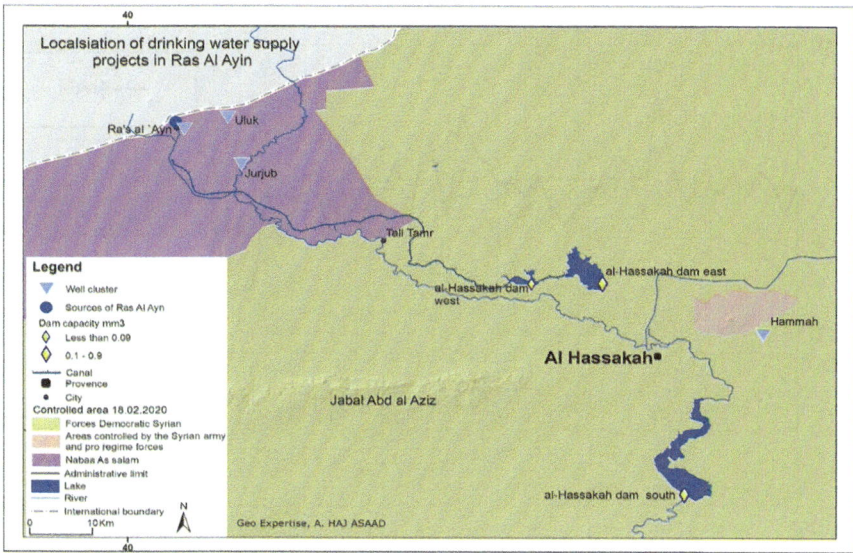

Fig. 13.5 Locations of various drinking water supply projects in the Ras Al Ayn area

water an unpleasant smell and taste. Therefore, it became neither suitable for drinking nor domestic use (Soumi and Danial 2010).

This decline in the quality and quantity of ground and surface water is the outcome of Syrian governments' development which focused most of their attention on the economic benefits of agricultural output while ignoring the environmental dimension. The extensive rechanneling of water for agriculture and consumption has increased electrical conductivity (the criteria used to assess salinity). The impact of Turkish use of groundwater in the upper basin on the decline in water levels in Ras Al Ayn springs cannot be ignored. Furthermore, the extraction and transportation of oil has contaminated water and soil since the physical networks used were too old and damaged, resulting in agricultural land pollution. We would like to point out that the expansion of private oil refineries in the aftermath of the Syrian conflict has increased pollution levels because these refineries discharge their waste into the valleys (Hindawi 2021).

Additionally, the dumping of untreated solid waste near water sources (such as the Qamishli landfill in the Al Hilaliyah area, and the Tall Abyad landfill in the Al Balikh sources) contributed significantly to water pollution, which is caused by the dumping of untreated sewage into natural watercourses and valleys, as well as into dam lakes, resulting in the contamination of both the surface and groundwater.

Under these circumstances, a study completed in 2000 by the General Company for Water Studies in Homs suggested forcibly pumping water from springs and wells from water carriers in Ras al Ayn to ensure a permanent water source for Al Hassakah governorate. To address the issue of the Al Khabur River's dryness and provide drinking water for the city of Al Hassakah and the villages located on the

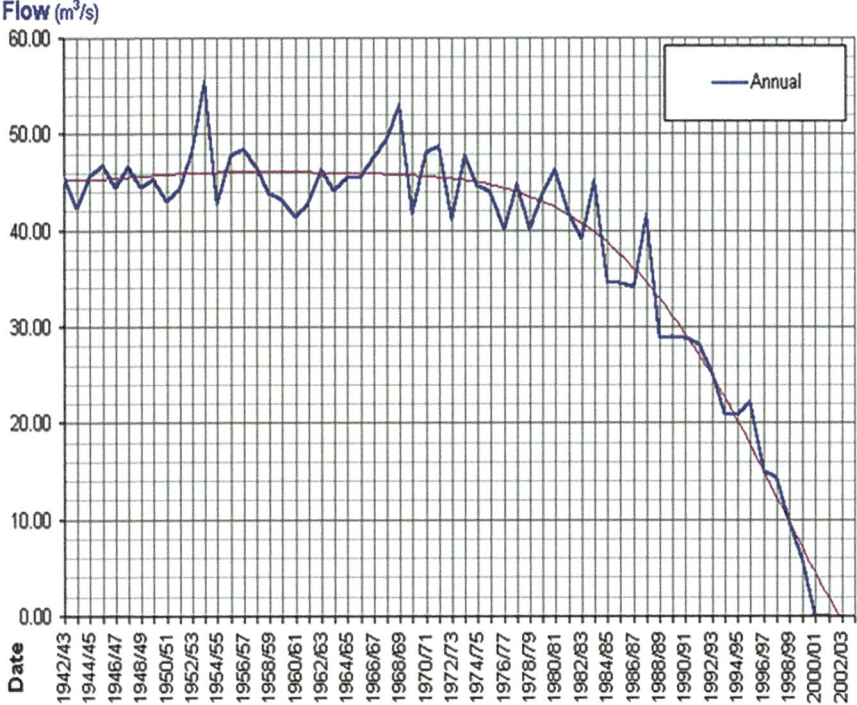

Fig. 13.6 Total changes in discharge of the Khabur springs in Ras al Ayn: 1942–2003

Table 13.3 Distribution of Wells

Number of wells	Location of wells	Supported areas
86	Ras al-Ain area	Supports main traction canal providing potable water for Al-Hasakah city and other cities and villages
30	Wadi-Jarjab area	Supports Khabur River

Khabur Valley, the Directorate of Water Resources in Al Hassakah drilled 116 wells near Ras Al Ayn and Wadi Jarjab (Table 13.3). These wells provided a flow of 15 to 20 m³/second. At the end of 2006, the productivity and discharge of these wells decreased to below 3 m³/s. Pumping operations from springs were also stopped as a result of damage to mechanical equipment due to flooding in the 2005–2006 season.

After that, two plans were developed to secure a water source for the 450,000 inhabitants of Al Hassakah and Tal Tamr and the surrounding rural areas. The first plan was adopted after Türkiye approved an Iraqi proposal to allow Syria to build a pumping station in Ayn Daiwar to draw Syria's share of the Tigris River water (1250 million m3/year) to Al-Hasakah, securing drinking water and irrigating 150,000 hectares of cropland. The plan was developed, and constructions were started, but it was stopped by the outbreak of the Syrian Revolution and the subsequent civil war.

The second plan was an emergency plan. It consisted of drilling 30 new artesian wells in the vicinity of Uluk village and transporting their water through pipes to the Tall Tamr and Al-Hassakah stations to secure water until the project of water drainage from the Tigris River to the al Khabur Basin could be implemented (Al-Hussein 2020).[7] 30 wells were drilled in Uluk village in 2010. 70,000 m^3 could be pumped per day from these wells (Fig. 13.5). The station also has a water tank with a capacity of 25,000 m^3. The water is pumped by 12 horizontal pumps through 70 km long transmission pipes to be delivered to the al Hammah water station, and from there to residential communities in Al Hassakah Governorate. There are currently 8 pumps in operation. Due to the war, ten wells were out of Service and some equipment and devices failed and were lost. Consequently, the amount of pumping from these wells was reduced to 50,000 m^3/day after. The plant relies on electricity from Tishreen Dam, which is controlled by the SDF.

Al-Hassakah, Tal Tamr, Ras Al Ayn city, as well as a large number of towns and villages, have become dependent on drinking water from the Uluk-Ras Al Ayn station. Some of them are supplied through networks that feed directly from the reservoirs of the al Hammah area. As for the rest of the rural areas and cities of Al-Shadaadah, Abu Rasin, and Al-Hawl, they are supplied with water through tanks, which are also supplied from the Al Hammah area. These cities are located in areas controlled by conflicting parties, which means that their water supply network crosses the boundaries of control. This situation was not unique to the Uluk-Ras Al Ayn station, but it is representative of the majority of water supply networks in Syria.

At the outbreak of the crisis, Al-Hassakah welcoming many displaced people. Al-Hassakah and its countryside have received a large number of displaced people from different parts of Syria during the conflict. Currently, the population dependent on the Uluk station is estimated at approximately 800,000 people. Al-Hawl, Arisha and Asho Kani camps are also supplied by tankers. Militarily, the Uluk-Ras Al Ayn drinking water pumping station and some of the cities and countryside benefiting from it are under the control of the National Army and the Turkish forces supporting it. However, most of the other cities and villages dependent on the station, such as Al Hassakah and Tall Tamr, are under the control of the SDF, except for the security zone, controlled by the regime forces and pro-regime forces.

It should be noted here that it is the civil agencies that carry out the technical management of water stations. The various military forces have preserved the governmental institutions and their structure while changing their names and references. Although the reference for water management and distribution in areas controlled by the National Army is the local council, the Water Resources Management Unit is responsible for water management in areas controlled by the

---

[7]It produces 630 megawatts, has a reservoir 155 km^2 in area, and a storage capacity of 1.9 billion m^3.

Table 13.4 Entities and institutions responsible for water management

Local military forces	Technical and administrative entities
Syrian democratic forces	Directorate of water resources and autonomous administration
National army	City Council of Ras al-Ayn
	Directorate of water resources in Ras al-Ayn
Regime forces	Ministry of water resources in Hasaka
	Directorate of water resources

autonomous administration. As for the security zone controlled by the Government of Damascus, the Water Resources Department of the Ministry of Water Resources has been preserved (Table 13.4). The disconnection of the city of Al-Hassakah and other towns and villages from water supplied by the Uluk-Ras Al Ayn station in the summer of 2020 was not an isolated or single event in Syria during the war. Most of the potable water stations were un-operational, whether by intentional or unintentional bombing, or by the intentional cutting off water to beneficiaries as a means to force residents to leave their villages, towns, and cities. In other words, water has been used as a tool of war.

However, the interruption of water supplies from the Uluk-Ras Al Ayn station has received unprecedented coverage by local and international media, as well as by activists who have mounted contradictory advocacy and mobilization campaigns in social media. Each party, depending on its affiliation and orientation, throws responsibility on the other to strip it of its credibility in the eyes of the local population, and tarnish its image at the local and international levels. While supporters of the SDF accuse Türkiye for the interruption, supporters of the National Army blame the SDF. The two parties have not ceased to accuse each other.

In the same context, some civil society organizations in northeastern Syria filed a complaint about the use of water resources for political purposes to five United Nations special rapporteurs. They were reporting on respect for human rights: the right to clean drinking water, to sanitation services, to adequate housing, the right to food, the right of every human being to the highest possible standard of physical and mental health, and human rights of internally displaced persons.

13.6.2 Energy Versus Water

The opposing local forces, represented by the National Army and the Syrian Democratic Forces, used rhetoric accusing the other of creating the conditions that forced the station's water pumping to stop. The National Army links the cessation of drinking water flowing from the Uluk station to a power failure caused by the SDF, because they control the energy facilities that supply the station with power from the Tishreen Dam, which they also control, forcing the Uluk station to shut down. The SDF justify their final power shutdown by the frequent violations perpetrated by certain of the National Army's components. According to the SDF, some groups of the National Army do not commit to operating the station for the sufficient and agreed-upon period required to provide the necessary quantity of water to Uluk station beneficiaries. According to the SDF, certain components of the National Army cut off water from time to time, despite the availability of energy required for running the station.

It accuses the National Army of supplying electricity to the villages and cities of the Nabaa as Salam area instead of supplying the pumping station. Therefore, the autonomous administration in Al Hassakah completely cut off electricity to the area until they can find a permanent solution to the dilemma. While the autonomous administration wants to provide electricity solely to the pumping station, the Ras al Ayn local council requests that electricity be provided to its service territory and that the Ras al Ayn and Tall Abyad regions be supplied with the equivalent of 30 mega-watts to ensure a sufficient quantity to meet the needs of residents and the Uluk station of electricity. The autonomous administration considers this amount to be excessive and requested that energy should be provided solely to the station. As a result, the autonomous administration in Al Hassakah directed that the amount of energy be reduced from 30 megawatts to 20 megawatts, and then to 12 megawatts until it was totally cut off. The Ras Al Ayn local council also states that the autonomous administration prevents the entry of fuel into the Salam spring area and points out that the SDF is using water as a tool for war by preventing the entry of fuel necessary to run pumping stations for irrigation and drinking water.

13.7   Water Station Management as a Tool to Enhance the Role of Political Actors

According to the regime, represented by the Directorate of Water Resources in Al Hassakah, both the autonomous administration in Al Hassakah and the local council in Ras Al Ayn have failed to provide water to the population. The directorate positions itself as an alternative to managing the station as a way of resolving the issue and ensuring the population's access to water. The local council in Ras al Ayn and the autonomous administration in Al Hassakah both opposed this proposal, claiming the regime's inability to provide water to the areas under their control. As major participants, Türkiye and Russia have been negotiating to solve the water shortage crisis. Russia has suggested that the regime's Water Resources Directorate can supervise the functioning and allow maintenance employees to enter the station. Russia's recommendation to allow the regime's technical presence in the area reveals its strategy of strengthening their ally's political presence in the Nabaa as Salam under the guise of technical assistance. In fact, the Russians and the regime are trying to use water as a tool to expand their political influence in that region, as its presence is still limited to the Security Zone in Al Hassakah and Qamishli. Control of water implies control of land and of political decision-making (Haj Asaad and Jaubert 2014).

13.7.1   Restructuring Tribal and Clan Alliances with Local and Regional Players

We have pointed out earlier that the northeastern region of Syria is made up of a large number of ethnicities and religions. They have maintained their identities despite the weak involvement of their sheikhs and leaders in managing public affairs of their communities due to the Baath Party's policy, which tried to undermine tribal and religious affiliations in favor of Baathist ideology. Despite the policy of rapprochement by former President Hafez al-Assad toward the families of tribal sheikhs to ease the pressure on them once in power, it only offered the sheikhs a narrow margin of operation in their relationship with their clan's members. Additionally, this margin could not exceed the limitations imposed by the regime. With the beginning of the withdrawal of government institutions from some areas, the tribal sheikhs' families began to regain some of their roles in representing their clans and tribe. But their positions and political orientations were scattered among the conflicting forces. Frequently, the clan or tribe's dispersal takes place inside the same chiefdom family. Regional and international powers have also sought to attract tribal families to their side. They were invited to the Russian military base in Humaymim-Latakia (Al Kanj 2019), to Tehran, and to Arab countries. They also organized a founding conference in December in Istanbul, and a conference in February of 2018 in Mersin, Türkiye.

The summer of 2020 witnessed a wave of assassinations in Northeastern Syria, the most impactful one targeted the Arab tribal sheikhs near Dayr az Zawr, controlled by the SDF. Sheikh Ali Suleiman Al-Weiss (from the al-Baqara tribe) was assassinated in the town of Al-Dahla in the countryside of Dayr az Zwar on June 31, 2020. The same fate befell Sheikh Mutashir Al-Hefl (an elder of the Al-Aqeedat tribe) in the town of al-Hawaij in the countryside of Dayr az Zawr on August 2, 2020. The SDF was accused once more of assassinating Arab leaders to terrorize opponents of their presence in Dayr az Zawr, ar Raqqa and the countryside around, and to single out the government. These assassinations provoked major protests against the SDF's presence in the Dayr az Zawr region, with the number of demonstrators steadily increasing and the protest area rapidly expanding.

The situation was exploited by tribal sheikhs who see the SDF as a threat to the Arab presence in the Jazeera. They mobilize their popular bases in the region, calling them to reject and resist the presence of the SDF in the regions of northeastern Syria. The majority of them sought refuge in Türkiye. However, a new perspective has emerged. Former ambassador Nawaf Al-Fares, who belongs to the Dulaim clan of the Al-Aqeedat tribe, which has a strong presence in northeastern Syria, particularly in the Dayr az Zawr districts, urged the clans to self-restraint.

While the ambassador's appeal was welcomed with enthusiasm in the tribal community, his invitation sparked debate in political circles, and some stakeholders and analysts have compared Al-Fares' views to that of the United Arab Emirates, where he resides and supports the SDF. It is the case of the defected Brigadier General of political security Nabil al-Dandal, (one of the sheikhs of a clan from the

Al-Aqeedat tribe) who currently lives in Zurich, stating in an audio message that spread on social media that the ambassador's position is irresponsible and in line with the SDF. While Sheikh Mahmoud Al-Murr agrees with the ambassador's position, he is concerned that the SDF will blame the tribes for terrorism and conduct a military operation that will expel the Arabs out of the region.

In the midst of these tensions, the step taken by the SDF to cut off electricity to the Uluk Station, leading to the cessation of water pumping, is an excuse to redirect the local community priorities and to damage Türkiye's image among the tribal grassroots. Furthermore, they tried to find a rift that can stop the rapprochement between the tribal sheikhs in Türkiye and their popular bases, complaining about SDF's actions, to dismantle their alliance and to prevent the creation of a strong axis hostile to the SDF presence in areas with a large majority of Arabs, while strengthening the axis of their allies. It is evident that the SDF seeks, whether by accident or intentionally, to use water distribution as a political tool to alter the existing power dynamics between the conflicting forces and rebuild them again for their interest.

13.8 Motivating Humanitarian Organizations to Work in Northeast Syria

The humanitarian aid coming to Syria is distributed over the three areas of control: northwestern Syria, northeastern Syria, and the area controlled by the regime forces. Humanitarian aid enters northwestern Syria from Türkiye, while non-UN humanitarian aid enters northeastern Syria from Iraq, from the Al-Yaarubiyah border crossing. The aid of the UN and its agencies enters through Damascus because it is not covered by Security Council Resolution No. 2165.[8] Aid is also coming through Damascus to areas controlled by regime forces. Therefore, there is competition among the three areas to maximize the benefits of the incoming aid to Syria. In this regard, the interruption of water supply to cities, rural areas, and camps reliant on the Uluk station in August, a time of high temperatures, revealed how vulnerable Syria's northeastern region is.

It has also been used as a means of advocacy by civil society organizations and the media for the need to increase humanitarian assistance in these regions. The cessation of water pumping from the Uluk station was a mechanism to disrupt the existing allocation of humanitarian aid between the three areas of control, and to restructure its distribution in a way that leads to increasing the share of aid to the

---

[8] Resolution 2165 (2014) Adopted by the Security Council: Decides that the United Nations humanitarian agencies and their implementing partners are authorized to use routes across conflict lines and the border crossings of Bab al-Salam, Bab al-Hawa, Al Yarubiyah and Al-Ramtha, in addition to those already in use, in order to ensure that humanitarian assistance, including medical and surgical supplies, reaches people in need throughout Syria through the most direct routes, with notification to the Syrian authorities, and to this end stresses the need for all border crossings to be used efficiently for United Nations humanitarian operation.

northeastern region. It is difficult to quantify the aid received, however, several humanitarian groups working on northwestern Syria have tried to establish offices in Iraq and work from there. Organizations are often established with names that differ from the parent organizations in order to secure their activities in diverse controlled areas.

In the same context, the autonomous administration urged humanitarian organizations and donors to help the northeastern region of Syria to secure its water independence, even partially from the Uluk station. Among the proposed projects is supporting the drilling and equipping of 50 wells in the al Hammah area. It should be noted here that the water quality in the al Hammah area is not suitable for drinking. As the population struggles to access a regular supply of affordable water (almost 30% of the salary of a government employee), and as the discourse of the conflicting parties hurling accusations at one another escalates, the actions around Uluk station only contribute to deepening the rift between society's components.

13.9   Envisioning a Way Forward

Finding solutions that adequately and permanently meet the population's water needs is one of the most important steps to achieving stability and for reducing internal and external migration. This issue is often approached technically. International or non-governmental organizations have often been accredited to manage water stations as a temporary solution, which requires understanding them from a socioeconomic and political perspective. The aim is to avoid making the issue of water access a tool of political settlements between conflicting parties (political and military) and a military weapon, as was the case during the conflict.

Therefore, it is necessary to lay the foundations for water resource management and facilities and the financing of their operation, which will ensure that their use is not politicized. Creating such an entity first requires a full knowledge of water resources, especially those that are or could be used for political purposes. Detailed hydrogeological, economic and social studies must be conducted on each source of water and their facilities to determine the nature of natural hazards and water use issues, and how policies are implemented. Then, mechanisms that are compatible with the natural and human environment surrounding the source need to be developed that can prevent water from being politicized and weaponized. In addition to assessing scenarios for how to prevent future water supply crises, it is also necessary to identify what roles international and local actors have, focusing on local civil society and encouraging its contribution to keeping water out of political bargaining, enhancing its ability to increase social cohesion instead.

The need for a detailed study of each water distribution network arises from the diversity of the Syrian population, where ethnic and religious affiliations differ from one area to another. It is therefore not possible to adopt a unified policy for all Syrian water sources and facilities, especially because of the long war Syria is still suffering.

13.10   Conclusion

The irrational policies of using water for irrigation and drinking purposes in the Ras Al Ayn Springs basin of the Jazeera in both Syria and Türkiye have led to a quantitative and qualitative deterioration of Water Resources. The flow of Ras Al Ayn Springs stopped in 2001, the Khabur River dried up, and the salinity of groundwater also increased. On the other hand, the policies adopted in the settlement of the population for political, military-security and ideological goals in the Jazeera since the beginning of the last century, have contributed to the construction of a conflictual social environment bound by a fragile social contract, which is soon torn apart to reignite conflict. Thus we witnessed the events of Qamishli in 2004, and then armed conflict between military forces erupt at the end of 2011, resulting in displacement of the population from all components of Jazeera society. This fueled tensions between the different social groups and contributed to the deepening of a feeling that living together was difficult, which weakened social cohesion.

Managing diversity in this environment is one of the current challenges; therefore, it is imperative to strive to accomplish it. Water can play an active role in this by enhancing social cohesion because of its characteristics as an essential element for everyone, because water is the basis of life and one of the gifts that increase the social capital (or in other words, the social relevance) of its actor. But water has been used—often during the ongoing conflict—as a weapon to achieve gains on the ground and to displace the population. Many drinking water supply and irrigation networks located on Syrian territory have been deliberately cut off, and water tanks and stations have been bombed because they are located in conflicting zones of military influence. The repeated interruption of drinking water from the Uluk Station since November 2019, can be classified as politicizing the use of water to achieve political gains for the competing parties, each related to the internal policy peculiar to the territories it controls, and each trying to affect the balance of power. The danger caused by the use of water as a military weapon or as a political tool is similar in its negative effects on society and the state.

The cutting of water supplies have exacerbated social tensions between the population components of Jazeera society. In addition, they have aggravated the water crisis and its high price. The price of a barrel of water (a barrel capacity of 200 liters) has become equivalent to 2% of the monthly income of an average civil servant, knowing that a person needs a minimum of 20 liters per day. Therefore, political and development actors in the early recovery and reconstruction phases, wishing to achieve socio-political stability and economic growth, should work on:

1. Finding a formula that contributes to both reconstruction and strengthening of social ties.
2. Avoiding mistakes in water management before and during the conflict.
3. Supporting the formation of independent committees consisted of technical and social experts to monitor the supply of drinking water, avoid the political use of water in the early recovery period, and reduce the resort to the use of water as an

instrument of blackmail to pass the policies of competing groups at the expense of the suffering of the population, thus undermining security and civil peace in the Reconstruction period.

References

Ababsa M (2009) La recomposition des allégeances tribales dans le Moyen-Euphrate Syrien (1958–2007). Études rurales 184:65–78
Al Jazeera News (2009) 1962 Census. Available at: https://lc.cx/pda-jZ
Al Kanj S (2019) How does each party in the Syrian Conflict use tribal influence? Chatham House. Available at: https://lc.cx/0JuyMn
Al-Hussein M (2020) What is the story of 'Ras Al Ayn Wells' and its relationship to the Al Hassakah water crisis? Economy. Available at: https://www.eqtsad.net/news/article/29522/
Balanche F (2018) Le nord-est Syrien, entre l'enclume du PKK et le marteau turc. Paper presented to the Secretariat of State for Migration, Federal Department of Justice and Police, Bern
Benhaim Y, Quesnay A (2016) L'espace politique kurde dans le conflit Syrien: intégration régionale et polarisation partisane. Confluences Méditerranée 3(98):75–87
Boris J (2010) Une ethnographie succincte de 'l'entre-deux kurde' au Moyen Âge. Études rurales 186:21–42
De Vaumas É (1952) Les grands travaux d'irrigation en Syrie. Annales de Géographie 328:471–473
Gibert A, Févret M (1953) La Djezireh syrienne et son réveil économique. Revue de géographie de Lyon 28(1):1–15. Lyon
Gorgas JT (2009) Repenser les nationalismes 'minoritaires': le nationalisme kurde en Irak et en Syrie durant la période des Mandats, entre tradition et modernité. A contrario 11(1):151–173
Haj Asaad AH, Chamali O (2016) Strengthening civil society organizations and improving access to water in northwestern Syria, Unpublished internal report
Haj Asaad AH, Jaubert R (2014) Geostrategic stakes and the impact of the conflict in the Orontes River basin. Confluences Méditerranée 89:173–184
Hindawi S (2021) Zero distance: east of the Euphrates. Al Jazeera New, Available at: https://www.youtube.com/watch?v=pTV0ZHCVvQo
Olson R (1989) The emergence of Kurdish nationalism and the sheikh Saïd rebellion, 1880–1925. University of Texas Press, Austin
Soumi G M, Danial D (2010) Resources crisis in the eastern region of the Syrian Arab Republic. Paper presented at the economic Tuesday symposium, Economics Society, Damascus
Syrian Republic, Central Bureau of Statistics (2010) Damascus
UN-Water (2013) What is water security? Infographic. Available at: https://www.unwater.org/publications/what-water-security-infographic

Chapter 14
Mining and the Local Dynamics of Territorial Governance in Southeast Morocco

Karen Rignall, Atman Aoui, and Moulay Ahmed el Amrani

14.1 Introduction

This chapter focuses on mining, which is currently framed not only as an important contributor to national development in Morocco and elsewhere, but is also a high-stakes contributor to energy transition initiatives because of the importance of rare earth materials and other key metals and minerals, such as copper and cobalt, in renewable energy commodity chains. In Morocco's newly published Plan Maroc Mines, for example, a central objective of the strategy for developing extractives is to "Positionner le secteur minier national comme pourvoyeur d'emplois, de prospérité et de richesses aux populations"[1] (Royaume du Maroc 2021a:12). Putting aside potential contributions to national GDP and state coffers, the idea that extraction supports local development or can be part of a sustainable development strategy has been shown to be true *nowhere*. This has been documented in the "resource curse" literature, which is based in the disciplines of economics and political science and has been critiqued in the literatures from critical geography and energy studies (Bebbington et al. 2008; Gilberthorpe and Rajak 2017; Ye et al. 2020). We are sympathetic to these critiques, which downplay histories of colonialism and

---

[1] To position the national mining sector as purveyor of jobs, prosperity, and wealth for populations.

K. Rignall (✉)
Community and Leadership Development Department, College of Agriculture,
Food and Environment, University of Kentucky, Lexington, KY, USA
e-mail: karen.rignall@uky.edu

A. Aoui
President of the Moroccan Association for the Promotion of Mediation,
Tinghir Province, Morocco

M. A. el Amrani
Civil Society Activist, Tinghir Province, Morocco

K. Darmame, E. Ross (eds.), *Local Governance and Development in Africa
and the Middle East*, Local and Urban Governance,
https://doi.org/10.1007/978-3-031-60657-1_14

post-independence imperialism, but we also see methodological utility to documenting what kinds of path dependence happen in extraction zones because of the long-term effects (the "resource curse") associated with extraction. How exactly, though, does extraction unfold in the context of local territorial governance, especially in the southeastern region of Drâa-Tafilalet? How do residents experience extraction as an economic, social, political, and environmental dynamic over which they usually have very little control? And how does extraction intersect with local governance, as rural communes attempt to meet the challenges of new constitutional mandates to assume greater decision-making authority and encourage citizen participation in local development processes?

These are big questions. For this chapter, we lay out a framework for analysis by presenting preliminary findings of a community-engaged research project comparing local experiences of copper mining in Bleida, Zagora with renewable energy in Midelt, Morocco that began in the fall of 2021. The project, led by the Association pour la Promotion de la Médiation au Maroc, explores local experiences of extraction and renewable energy with the goal of supporting the democratization of knowledge and greater citizen participation in decision-making around extraction. This includes participatory action research through community-based and documentary research methods. We are especially interested in how large projects like mines or solar energy plants influence the relations between local elected officials, the state, companies, and residents. We are also developing a method for assessing the fiscal and economic impacts of mining at the local level, both in the narrow sense of revenues and costs to local governments and in the broader sense of resource dynamics that affect access to natural resources beyond the extracted metal or mineral. Research activities are focused in the communes of Bleida, the site of the Managem-owned copper mine, as well as MiBladen, Zaida, and the city of Midelt, where a large-scale solar installation is currently being developed.

Parallel to our community engaged methods, we are conducting an analysis of Moroccan mining and bureaucratic procedures so that we can "translate" them for residents who do not have access to this information. In what follows, we present our first iteration of this bureaucratic and juridical analysis, which will be revised and updated in conversation with residents and as the context changes: laws and procedures regarding mining operations and impacts that are relevant for understanding local territorial governance. The starting point for this analysis is the revised mining code of 2015 and the Natural Resource Governance Institute's (NRGI) annual analysis of governmental transparency in the extractives sector; the global accountability organization conducted this analysis in 2021 for only 18 countries, which speaks to the importance of extraction in Morocco and Morocco's global importance for the extractives sector, especially for its dominance in phosphate production (NRGI 2022). Morocco scored low overall (a rank of "weak" in the 5-category system of good, satisfactory, weak, poor, and failing) and went down in its score since it was last assessed in 2017, largely because of Moroccan government moves to restrict transparency in reporting on phosphates. While there are limitations in summarizing complex phenomena such as transparency and accountability in extractives in an index format, the Natural Resources Governance Institute analysis serves as a foundation for our own project that we then ground truth through our ethnographic and other documentary research.

14.2   Moroccan Extraction in Context

The laws and regulations governing mining are part of a larger Moroccan government project to intensify resource extraction, especially in the non-phosphate mining sector. This project is summarized in the *Plan Maroc Mines*, which follows the template for similar government initiatives as the *Plan Maroc Vert* (agricultural development), *Plan Maroc Bleu* (water management), and *Plan Solaire* (solar energy). The contextual timeline offered in the Plan Maroc Mines strategy document refers to "before 2019" only in terms of the "évolution technologique de l'industrie minière à l'échelle international qui ouvre la voie aux opportunités offertes par le secteur minier national"[2] (Royaume du Maroc 2021a, b: 8). New markets and technological opportunities internationally are, in fact, opening resource frontiers for Moroccan mining that make exploiting new deposits and extending the lives of old mines more economically feasible than ever before. A longer view, though, reveals the extent to which extraction has played a key role in the development (or gaps in development) of infrastructure and strategies of control in the rural southeast from the earliest days of French colonialism (1912–1956), even before the French secured full military control of mining sites such as Ouarzazate, Tinghir, and Zagora, in the Draa-Tafilalelt region.

    The legal architecture used to facilitate European extraction of Moroccan metals and minerals relied on outright expropriation or on land tenure laws created to support European expropriation. Laws were formulated after the fact to control speculative encroachments especially by French and German companies in partnership with the Moroccan *caids*[3] notorious for their repressive rule (Empire Chérifien/ Protectorat de la République Française au Maroc 1914). Only in 1951, the waning years of the Protectorate, was the mining code promulgated at the beginning of French rule revised, and it had few elements conventionally accepted as necessary for contemporary regulatory frameworks, such as free, prior, and informed consent; environmental and social impact assessments; compliance and enforcement mechanisms; and decommissioning regulations. Our ongoing archival research at the Archives du Maroc reveals how exploration and exploitation of Morocco's sub-surface shaped the economic and political role of the southeast in the French colonial project.

---

[2] The technological evolution of the mining industry at the global scale which creates opportunities for the national mining sector.

[3] Caids: traditional authority figures, often heads of powerful tribal fractions or confederations, whose authority over rural and pastoral areas and populations was recognized by the Moroccan State.

Mines established in this speculative and "extra-judicial" environment were subsequently brought into the domain of the state or handed over to private companies that are now the backbone of Morocco's extractive sector, such as Managem, the private company with majority royal interest that dominates the non-phosphate mining economy. While our research on the history of extraction in the southeast is ongoing, evidence shows how these peripheral regions were not, in fact, dismissed by French colonial policy as "le Maroc inutile."[4] This was the famous phrase that France's first Resident General in Morocco, Hubert Lyautey, used to justify colonial policies that devoted infrastructural and other development efforts to the littoral plains and fertile agricultural regions open to European colonization. Rather, the southeastern periphery was "utile" in very specific ways, as road construction and other investment were directed to prospecting and exporting valuable minerals and metals. The nature of extraction enables enclave investments that simultaneously export wealth while depleting or neglecting the existing wealth and lifeways of extraction zones.

14.3   Transparency and Governance in Contemporary Extraction

This long and contested history forms the background to contemporary transparency issues that have impacts for territorial governance in communes with extraction. The current legal and bureaucratic framework for mining is elaborated primarily in the 2015 reform of the mining code of 1951. The recent reforms are intended to bring Moroccan extraction law into line with international norms, but the preamble is also clear that the law aims to provide incentives and bureaucratically facilitate an acceleration in extractive activity, primarily by the private sector. In lieu of a detailed analysis of the reforms (see El Atillah et al. 2018), we highlight here the legal and bureaucratic dimensions of extractives governance at the national scale that are relevant to how local communes govern their territory. The three domains that are emerging as important for residents and other stakeholders in our comparative project are:

1. **Resource ownership, access, and use**: This includes who has jurisdiction over what can happen at the local level or in the commune's territory and how local government dynamics respond to conflicts or ambiguities regarding how resources are managed and transferred;
2. **Fiscal and economic impacts**: This domain relates to decentralization initiatives and the uneven way advanced regionalization in Morocco intersects with the centralized control of the extractives sector and the relationship between the state and private mining companies;

---

[4]"Useless Morocco."

3. **Environmental impacts and other transparency/compliance issues**: The legal
   framework for environmental assessments and compliance has been elaborated
   somewhat in recent years but the gap between law, procedure, and practice is
   notable in this domain.

Running through these domains are issues of transparency and whether residents are
being meaningfully consulted as projects are being developed. Extractives law in
Morocco integrates few of the international conventions for free, prior, and informed
consent that have become the norm for international finance institutions and watch-
dog groups (NRGI 2022). Even these processes have been roundly criticized for not
allowing for meaningful participation and decision-making of affected populations
(Leifsen et al. 2017). There are other mechanisms in the Moroccan constitution and
territorial governance that allow citizens to make claims about democratic participation and equity; the short preamble to Articles 168–171 of the constitution
specifically mentions "the promotion of sustainable, human development and participatory democracy" (Royaume du Maroc 2021a, b: 59). However, these *instances*
have yet to be used to successfully open the "invited spaces"—the spaces of
decision-making that are not open to broad participation—of extractives governance
(Hunjan and Petit 2011). The consultative processes established by the 2011 constitution are beyond the scope of this chapter.

14.4   Resource Ownership, Access, and Use

As in most countries around the world, the Moroccan state claims exclusive ownership of sub-surface rights. The 2015 mining code elaborates on these ownership
rights, noting that the state assumes ownership of all mines, not just minerals or the
rights to the sub-surface (Royaume du Maroc 2015). That private companies receive
permits for exploitation does not mean that they own the mine itself, but rather that
they are licensed to exploit the resources in that mine. Given the levels of investment
required to prospect and exploit the sub-surface, legal guarantees about the right of a
private company to the long-term benefit of their investment are essential to the
hybrid public-private nature of the extractives sector in Morocco. The result at the
local level is a blurring of boundaries between the state and private companies in
terms of who has the right and responsibility of guaranteeing rights to exploitation
and managing dissent, which in Bleida and at other mines has taken the form of
demonstrations, direct claims to local state officials, social media campaigns, and
other press coverage, among others (El Kahlaoui and Bogaert 2019).

Claims about ownership, access, and use of resources therefore focus less on who
owns the copper in Bleida, cobalt in Bouazzer, or silver in Imider than on who can
reap the benefits from those resources and how (Aoui et al. 2020). One change in the
2015 mining code with potentially significant impact in this context is the rule that in
order to receive a permit, an applicant must be incorporated as a company, not an
individual person. This has implications for the unique situation in Morocco

whereby individuals and small-scale operators have been formally institutionalized as part of a network (CADETEF) that retains exploitation rights over a region with great extractive potential (El Atillah et al. 2018). The goal of opening this region to more corporate interests means that the rule requiring incorporation could have the effect of undermining artisanal or small-scale mining in favor of corporate concentration, funneling the economic benefits further upward in the mining value chain. The requirement that a permit applicant constitute a "moral person" (or is incorporated) and have the necessary technical and financial capacity to operationalize the permit may help to mitigate speculative pressures or alternately, use an ambiguous standard to crowd out local actors.

In addition to concerns about who can apply for permits and reap the benefit of mining, recent changes in mining law have implications for how local communes exert control over the territory in their jurisdiction. Law 33–13 of 2015 loosens restrictions on how much land can be authorized for exploration and exploitation, from 100 square kilometers previously to 600, with the possibility of one applicant receiving 4 adjacent permits, for a total of 2400 square kilometers (El Atillah et al. 2018: 10). The right to extract from the sub-surface can therefore take significant tracts of land out of the control of local actors: from collective owners to users and the commune authorities charged with creating *plans d'aménagement* (development plans). With increased investments in commercial agriculture in the collectively owned steppe of the southeastern provinces, there is increased competition in land uses and water consumption between agricultural concerns and mining activity. In Bleida, pressure on land because of the expansion of watermelon production for export was cited to us as a more pressing concern than mining activity, but each unfolds in relation to the other.

These competing land uses raise the question of who controls resource allocation between diverse sectors and who secures sustained access to those resources. What regulatory framework governs these competing uses? The permitting process is highly centralized in Morocco, with the National Office of Hydrocarbons and Mines (ONHYM) holding exclusive authority and permitting conducted on a first come, first served basis, as opposed to offering permits through a call for proposals for developing a given resource (NRGI 2022). The permitting process includes no process for free, prior, or informed consent for residents that enables input into the decision-making process, and no appeals mechanism. Although the ONHYM website offers a clear organizational chart of the permitting process, the Plan Maroc Mines includes proposed changes to streamline and incentivize investment from domestic and foreign companies (ONHYM 2022; Royaume du Maroc 2021a, b). In any case, law 33–13 does not require the licensing authority to disclose the full list of firms that hold licenses and permits, and no such comprehensive list has been published.

Delving into the full cycle of permitting for exploration, exploitation, compliance, and decommissioning reveals ambiguities and even contradictions in jurisdictional authority that complicate territorial governance for communes trying to conduct comprehensive economic development planning. In the commune development planning process, for example, there is a procedure for public participation; residents have the right to give their feedback on the draft plan, even if this right is little known and rarely invoked (project partners did oversee such a process in a Tinghir commune that showed the possibility for mobilizing commune residents). This process provides an entrée for communes to impact planning processes at the provincial council level. However, with no recourse for residents or commune governments to impact permitting decisions for extraction, territorial governance at the local level is inherently circumscribed. The lack of an overarching framework for resource and territorial management means that commune governments and different actors—notables, corporate actors, the Ministry of Interior, customary leaders, and residents seeking a voice in these deliberations—are negotiating, arbitrating disputes, or imposing solutions from above. Such ambiguities of jurisdiction extend beyond the permitting process to who holds authority for ensuring operational compliance with permits, licenses, contracts, and other obligations. Article 94 of Law 33–13 notes that agents of the administration have authority over compliance but the law does not specify what to what entity the term "administration" refers (NRGI 2022; Royaume du Maroc 2015). It is also difficult for observers to determine if companies are compliant with their fiscal operations: the national tax authority is not required to audit the private extractive companies (Royaume du Maroc 2004).

The next section examines these fiscal dimensions in more detail. Here, we conclude with our argument that the legal framework for ownership of sub-surface rights intersects with the overall structure of the mining sector as a legacy from the colonial period. The principal instituted from even before the Protectorate (a period of rampant speculation and aggressive resource prospecting that the Treaty of Algeciras was intended to routinize) is that of "first come, first served" for permitting of exploration and exploitation (Rivet 1979; Royaume du Maroc 2013). The state does not make sub-surface reserves public or issue calls for proposals from potential licensees. Any given local actor, from collective landowners to commune governments, are therefore left guessing about what resources may sit below the surface and what entity may propose to exploit them, making long-term development planning a complicated, potentially contentious challenge.

14.5 Fiscal and Economic Impacts

One of the primary issues for local governments in extraction zones is the question of how much and in what way they benefit from or incur costs because of mining operations. Demands for good paying jobs are one of the most pressing claims in mining sites and other large investment projects like solar energy in Morocco (and around the world) but the fiscal and economic impacts of mining extend beyond the household to the ways local communes receive and manage public revenues associated with those operations (Cantoni and Rignall 2019; Rignall 2016). The major concern here is what portion of tax or valued-added revenues return to the communes. Historically, these revenues went exclusively to the regions' general funds, with the effect of funneling economic benefits to higher levels of territorial governance. With the 2015 mining code, this formula was changed to allow for 50% of those revenues to return to the commune where primary operations occur. For a mine like Bleida, with operations completely within the borders of one commune, this revenue allocation should be straightforward. However, our research has shown that officials' understanding and implementation of the new allocation procedures is uneven. This is one domain where the democratization of knowledge—an objective of our action research process—is particularly important. A company like Managem, the owner of the copper mine at Bleida and other prominent mines in the region, has the resources and expertise to examine their legal obligations in detail. Company officials do not have the obligation nor do they take the initiative to inform commune governments or other local actors of receipts or legal provisions beyond the basic requirements imposed on the company. This can result in Managem paying rent on collectively owned land or other fees, for example, but those funds sitting in accounts unused or unallocated because the receiving government entities are unclear about their jurisdiction over those funds. To date, the commune of Bleida has not received any royalties or revenues according to the scheme outlined in the 2015 mining code reform; interviews also surfaced unconfirmed assertions that Managem's lease payments on the collectively owned land in Bleida are not being accepted or used by the Ministry of Interior because of confusion about whether the latter is authorized to accept payments for sub-surface rights. Importantly, then, analyzing the fiscal and economic impacts involves more that assessing direct employment or investment in the commune coming from mining but also examining the full range of revenues returning to the mining site and the kinds of projects they enable or foreclose.

Part of the confusion surrounding when and how the 2015 rule reapportioning taxes to include communes as well as the region may be that the *loi d'application* (implementing law) of 2016 only addresses some provisions of the 2015 mining code, especially those relating to the approval or termination of mining titles (Royaume du Maroc 2016). We are in the process of verifying the status of the revenue allocation rule as there are wide ranging opinions from various sources about the status of this aspect of the law. In general, public and local officials have little understanding about fiscal regulations regarding mining, in part because there is no centralized place where these are made available in accessible language and because they are spread across numerous legal codes. The basic regulation according to the financial law of 2021 is that mining firms must pay taxes on mining exploration permits and operating permits, as well as a transfer tax (Royaume du Maroc 2021a, b). It remains a challenge, however, to track the range of revenues that are allocated at different levels of territorial governance. Even though the *code géneral des impots* (general tax code) outlines the income tax rates applicable to extraction companies, according to the NRGI transparency assessment, the Moroccan government does not publish data on the value of fiscal receipts and payments from these companies, nor is there a rule specifying how royalty rates are calculated (NRGI 2022). And while the value-added tax is divided between regions and communes and therefore should be available to stakeholders, this is only one form of revenue from mining, and critical reports like the NRGI have noted an overall lack of transparency regarding revenue flows and fiscal processes. This opacity occurs in the way the laws are designed and in the gap between practice and policy. Therefore, while the government projects revenues from phosphates and from mining firms, we do not have a way to comprehensively learn what those revenues actually are. The High Commission for Planning "key figures" document with statistics about the Moroccan economy, and the summarized data on the OHNYM website only offer a simplified and synthesized assessment (Haut-Commisariat au Plan 2021). Each government agency with some link to the extractives sector has its own internal processes and deals with the specificities of each mining company differently, further complicating transparency around extraction at all levels of government.

Beyond these fiscal issues, it is difficult to access clear and comprehensive data on mining production and the state of prospection efforts. At the local level, which companies are present is not a mystery, but how local production figures into corporate strategies for future exploration and growth and the broader context for mining are difficult to determine. According to the Mining Code of 2015, the Moroccan government is not required to and does not disclose mining production data in a detailed format. The Ministry of Energy Transition and Sustainable Development offers an aggregated assessment of the sector, and publicly listed companies like Managem issue annual shareholder reports (Royaume du Maroc 2019). Understanding what current production is and then what reserves exist and where, however, is not publicly available information; piecing together what data do exist requires expert knowledge of the sector and how to read financial reports. Coupled with the lack of transparency of how many permits there are, who owns them, and their details, this makes local knowledge about mining difficult (see, for example, a general map: https://mining.onhym.com/en/mining-projects-map and list of partners: https://mining.onhym.com/en/partners). Given the "first come, first served" system for applying for exploration and mining permits, commune governments and their residents have no way of knowing how their territory figures in corporate strategies for growth or the reserves in their boundaries so that they can plan for maximizing the economic potential of their resources or conduct systematic planning that places extraction in relation to other, perhaps more sustainable and equitable economic activities.

14.6   Environmental Impacts and Other Transparency and Compliance Issues

As with the fiscal and permitting dimensions of extraction, the legal framework for assessing and minimizing the environmental impacts of mining falls short of international best practices (NRGI 2022). Beyond the formal rules, however, a key issue of concern to residents and local collectivities is the gap between policy and practice as well as challenges of tailoring centralized, national laws to local contexts and territorial governance. Pressure from international finance organizations and a concerted effort to burnish the environmental credentials of the Moroccan government have led to a suite of environmental protection laws in recent years. Law 12-03, for example, requires the government to publicly disclose the results of environmental impact assessments, consult key stakeholders, and publicly detail the whole EIA process while another dahir (2-04-563) outlines the procedures for environmental impact studies (Royaume du Maroc 2008a, b, c). Despite procedures that appear to conform to international standards, neither the Moroccan government nor extractive companies make EIAs publicly available, though, as one elected official noted in Bleida, individual inquiries to the companies may receive favorable response. Environmental impact standards also do not address legacy projects, such as the Bleida mine, that began production before the laws went into effect. Formally adhering to standards for consultation or public disclosure also begs the question of how such consultations or disclosures unfold: when and where do they occur? Who is invited? Who receives information about it? Disclosures detailed in documentation to international funders may not have reached most residents or involved the full range of stakeholders, in part because of the format and language of such consultations. Other issues relating to impact assessments and compliance regard law 12-03's requirements that companies must prepare a plan to compensate and mitigate the negative impacts prior to development, but there is no public disclosure of these plans nor any elaboration of enforcement mechanisms. Similarly, there is no publicly accessible or routinized mechanism for residents or local officials to seek redress for infractions; claims and complaints are handled informally, often through direct entreaties to companies and/or Ministry of Interior officials. Such entreaties have been met in the past with repressive measures, such as arrests and other attempts at containing protest (Aoui et al. 2020; Bogaert 2016).

14.7 Conclusion

This analysis of how the national governing framework for extraction sets the terms for territorial governance at the local commune level in Morocco focuses largely on the juridical framework and supporting procedural or implementation directives. There are contradictions inherent in a highly centralized and yet at the same time skeletal legal framework that leaves many aspects of jurisdiction, accountability, and transparency unelaborated. The challenges for commune officials and local civil society activists to act on their own development priorities in an environment of limited knowledge and decision-making power are formidable. In our field-based interviews with residents and other local stakeholders around the Bleida mine, we have seen the effects limited knowledge about extraction law and procedures, from confusion about whether the commune should be receiving a share of tax revenues (it should, but disbursements have not yet been made as of this writing), whether permitting procedures for expanded mine activity are being followed, and how to determine whether environmental impacts, especially to water, have been documented in a robust and objective way, among others. As we continue both our field-based and documentary research, questions remain how about the legal framework can facilitate or hinder territorial governance. In this context, the broad objective of democratizing knowledge about extraction's history and legal framework can offer local actors additional tools for participating in decision-making around extraction projects and fulfilling their own development priorities in concertation with state agencies as well as the mining company.

Democratizing knowledge about extraction, however, has to involve more than collating the suite of laws and implementing regulations and making them available to officials, activists, and residents. Determining how to use legal instruments to make claims and otherwise work towards accountability in mining operations also requires a savvy *translation* of those instruments that outlines how people not versed in high-level policymaking can still use them for popular benefit. There are examples from environmental justice activism around the world that show the perils and possibilities of using legal and administrative tools to advocate for social change or enhanced local governance capacity (Kirsch 2014). Using existing law to ask for transparent reporting and accountability will not, in and of itself, redress the rural southeast's economic and political marginalization, which is the product of multiple processes and a long historical arc. But the project of clarifying what mining companies' legal obligations are and how the state can enforce those obligations is an important step in supporting local collectivities' aspirations to design development plans suited to their context. The starting point for this kind of analysis and outreach strategy is what do local actors and institutions foreground as their questions and priorities. What are the claims residents are making and how do these claims speak to the root causes for the region's marginalization? Local actors may not necessarily articulate transparency and accountability in national extractives governance as their primary objective, but in examining the claims they do make, such transparency and accountability in the elaboration of laws may be a necessary first step. The second step, then, would be how these laws and procedures take shape in the actual lived experience of residents and in mining operations.

There is an inevitable gap between the legal framework for extraction and its implementation, especially in regions such as the Moroccan southeast that have traditionally been sidelined in national development priorities and otherwise faced political repression. But our analysis cannot stop at documenting this gap. Beyond the question of whether the state is executing its own laws is whether those laws are substantively fair and just. With its roots in colonial policies explicitly aimed at conquest and expropriation, the legal framework for mining is arguably unsuited to a process of deliberative democracy whereby residents and local governments are involved in deciding how extraction can proceed with minimal harm and redistributed benefits, or if it should proceed at all. While it may not be realistic to expect that civil society actors or local government officials can prompt a rethinking of the equity impacts of mining at the national scale, a historical awareness that justice was never a factor in extractives governance can serve as a basis for making more expansive claims. Efforts to democratize knowledge about extraction can support social movement mobilizing but it can also be a more modest and practical tool for empowering local collectivities to more effectively target their governing approach to their local context, in Morocco and beyond.

References

Aoui A, El Amrani MA, Rignall K (2020) Global aspirations and local realities of solar energy in Morocco. Middle East Research and Information Project

Bebbington A, Hinojosa L, Bebbington DH, Burneo ML, Warnaars X (2008) Contention and ambiguity: mining and the possibilities of development. Dev Chang 39(6):887–914

Bogaert K (2016) Imider vs. COP22: understanding climate justice from Morocco's peripheries. Jadaliyya Nov 21. http://www.jadaliyya.com/pages/index/25517/imider-vs.-cop22_understanding-climate-justice-fro

Cantoni R, Rignall K (2019) Kingdom of the sun: a critical, multiscalar analysis of Morocco's solar energy strategy. Energy Res Soc Sci 51:20–31

El Atillah A, Souhassou M, El Morjani Z (2018) Le cadre législatif de l'exploration et la recherche minière au Maroc entre le Dahir de 1951 et la loi 33-13. Int Rev Econ Manag Law Res 1(1):1–20

El Kahlaoui S, Bogaert K (2019) Politiser le regard sur les marges. Le cas du mouvement 'sur la voie 96' d'Imider. L'Année du Maghreb 21:181–191

Empire Chérifien/Protectorat de la République Française au Maroc (1914) Dahir portant réglementation pour la recherche et l'exploitation des Mines dans la zone du Protectorat français de l'Empire Chérifien. Bulletin Officiel Édition Française 3(66):55–63

Gilberthorpe E, Rajak D (2017) The anthropology of extraction: critical perspectives on the resource curse. J Dev Stud 53(2):186–204

Hunjan R, Petit J (2011) Power: a practical guide for facilitating change. Cambridge United Kingdom Trust, Fife

Kirsch S (2014) Mining capitalism: the relationship between corporations and their critics. University of California Press, Berkeley

Leifsen E, Gustafsson M-T, Guzmán-Gallegos MA, Shcilling-Vacaflor A (2017) New mechanisms of participation in extractive governance: between technologies of governance and resistance work. Third World Q 38(5):1043–1057

Li F (2015) Unearthing conflict: corporate mining, activism, and expertise in Peru. Duke University Press, Durham

National Office of Hydrocarbons and Mines (ONHYM) (2022) Mining Exploration Partnership. https://mining.onhym.com/en/licensing-process-mining. Accessed 19 Oct 2022

Natural Resources Governance Institute (NRGI) (2022) 2021 Resource Governance Index: Morocco (Mining). https://resourcegovernance.org/sites/default/files/documents/2021_rgi_morocco_mining_profile.pdf. Accessed 29 Sept 2022

Rignall K (2016) Solar power, state power, and the politics of energy transition in Pre-Saharan Morocco. Environ Plan A(48):540–557

Rivet D (1979) Mines et politiques au Maroc, 1907-1914 (D'après les archives du Quai d'Orsay). Revue d'historie moderne et contemporaine 26(4):549–578

Royaume du Maroc (2004) Décret n° 2-02-121 du 24 chaoual 1424 relatif aux contrôleurs d'Etat, commissaires du gouvernement et trésoriers payeurs auprès des entreprises publiques et autres organismes. Bulletin Officiel, January 1, Rabat

Royaume du Maroc (2008a) Loi 12-03 sur les études d'impact sur l'Environnement. Bulletin Officiel, Novembre 20, Rabat

Royaume du Maroc (2008b) Décret n° 2-04-564 du 5 kaada 1429 (4 novembre 2008) fixant les modalités d'organisation et de déroulement de l'enquête publique relative aux projets soumis aux études d'impact sur l'environnement. Bulletin Officiel, Novembre 20, Rabat

Royaume du Maroc (2008c) Décret n° 2-04-563 du 5 kaada 1429 relatif aux attributions et au fonctionnement du comité national et des comités régionaux des études d'impact sur l'environnement. Bulletin Officiel, Novembre 20, Rabat

Royaume du Maroc (2013) Note de présentation du projet de loi n° 33-13 relative aux mines. In: Ministère de l'Energie, des Mines, de l'Eau et de l'Environnement, Département de l'Energie et des Mines, Rabat

Royaume du Maroc (2015) Dahir n° 1-15-76 du 14 ramadan 1436 (1er juillet 2015) portant promulgation de la loi n°33-13 relative aux mines. Bulletin Officiel, Rabat

Royaume du Maroc (2016) Décret n° 2-15-807 du 12 rejeb 1437 (20 avril 2016) pris pour l'application des dispositions de la loi n° 33-13 relative aux mines portant sur la procédure d'octroi des titres miniers. Bulletin Officiel, July 21, Rabat

Royaume du Maroc (2019) Indicateurs clés. https://www.mem.gov.ma/Pages/secteur.aspx?e=7. Accessed 19 Oct 2022

Royaume du Maroc (2021a) Plan Maroc Mines 2021–2030: Vers un modèle compétitif à l'horizon 2030 œuvrant pour une industrialisation intégrée et une croissance durable. Ministère de l'Énergie, des Mines, de l'Eau et de l'Environnement, Rabat

Royaume du Maroc (2021b) *Dahir n° 1-21-115* du 5 joumada *I* 1443 (10 décembre 2021*)* portant promulgation de la *loi* de *finances (76)21 pour l'année budgétaire* 2022. Bulletin Officiel, Decembre 20, Rabat

Royaume du Maroc, Haut-Commisariat au Plan (2021) Le Maroc en chiffres. Haut-Commissariat au Plan, Rabat

Ye J, van der Ploeg D, Schneider JS, Shanin T (2020) The incursions of Extractivism: moving from dispersed places to global capitalism. J Peasant Stud 47(1):155–183

Chapter 15
An Illustrative Model of (Non) Local Governance: Morocco's Coastal Zones

Samira Idllalène

15.1 Introduction

Coastal regions are home to a large portion of the world's population, along with some of the worst territorial governance issues. In Morocco, the coastal zone plays an ever increasing role in the country's economy. In fact, more than 50% of industry is located along the Atlantic Seaboard (Laouina 2016). By adhering to the SDGs and the blue economy paradigm, Morocco is focusing more on its coastal zones (CESE 2018). The Coastal Act was enacted in July 2015. It brought a new approach for coastal zone governance. This law defines the coastal zone as extending landward to the maritime public domain[1] and to wetlands, lagoons and dunes and seaward to the territorial sea (article 2-1).

The evolution of the legislation on coastal management has coincided with that of local governance.[2] However, on the ground, the principle of decentralization is barely applied to coastal management. The Coastal Act establishes a coastal

---

[1] The Coastal Act refers to the Dahir of 1st July 1914 on the Public Domain, which defines the maritime public domain as: "the shore of the sea at the the highest tides, as well as an area extending six meters above this limit, natural harbours, ports, harbours and their outbuildings, lighthouses, lamps, beacons and generally all the works intended for the lighting and for the marking of coasts and their outbuildings". Official Bulletin #89, 10 July 1914 (translation added).

[2] The Coastal Act was issued on 16 July 2015 (by the Dahir #1-15-87) and published in the Official Bulletin #6404 of 15 October 2015, p. 3746. At the same time, three main Acts on decentralization were enacted: namely the Organic Law #14.111 on Regions; the Organic Law #14.112 on Prefectures and Provinces and the Organic Law #14.113 on Communes (Municipalities).

S. Idllalène (✉)
Polydisciplinary Faculty, Law Studies Department, Cadi Ayyad University, Safi, Morocco

Faculty of Legal, Economic and Social Sciences, Laboratoire de recherche sur la coopération internationale pour le développement (LRCID), Marrakech, Morocco
e-mail: s.idllalene@uca.ac.ma

© The Author(s), under exclusive license to Springer Nature Switzerland AG 2024
K. Darmame, E. Ross (eds.), *Local Governance and Development in Africa and the Middle East*, Local and Urban Governance,
https://doi.org/10.1007/978-3-031-60657-1_15

planning system based on the "integrated coastal zone management" (ICZM) approach. ICZM is challenging for both local and national authorities. One of the most important challenges is the limited powers of these authorities in both their content and scope. This contradicts the principles of coastal governance as enshrined in international law (ICZM Protocol, SDGs, bleu economy...). In addition, it disregards the good governance approach promoted by the Constitution (Title XII).

As an area of interaction between land and sea, the coastal zone raises enormous challenges for decision-makers. For example, it is impossible to manage the land without taking into account the management of the sea and vice versa. Often coastal management is accomplished selectively and partially. This has an impact on coastal sustainability.

This chapter will address the challenges that the ICZM represents for both local authorities as well as the Moroccan government. First, it starts by addressing some important legal principles of coastal governance. Secondly, it explains the legal and institutional framework of coastal planning in Morocco. Thirdly, it analyzes the role of coastal stakeholders in coastal governance through a few case studies. It concludes by suggesting some changes to coastal governance in Morocco.

15.2   The Coastal Zone, a Valuable Space

A number of strategies and development plans have regulated the management and development of Morocco's coastal zone. These include: *Halieutis* (for the fisheries and aquaculture sector),[3] the *Plan Azur* (for beach tourism),[4] the *Plan Emergence* (for renewable energies and port infrastructure) (Amine 2016; Lmariouh & Askour 2021). As a rich and vulnerable ecosystem, the coastal zone is also covered by environmental protection legislation (Idllalène and Masski 2015). The coastal zone offers a multitude of ecosystem services and is coveted by a number of important economic sectors (tourism, maritime trade, aquaculture, fisheries, desalinization plants, mining, etc.) (PACC 2022).

Economic activities and the strategies that frame them can have huge environmental impacts on the coastal zone (Nakhli 2010; Mansoum 2016; Laouina 2016). These impacts are exacerbated by the effects of climate change (sea level rise, ocean acidification). Therefore, due to the competition for space in the coastal zones, these areas have particular characteristics and needs that require a holistic approach. The

---

[3]*Halieutis*: Strategy on Competiveness and Sustainability of the Fishery Sector was adopted in 2009 by the Ministry of Fisheries. The National Agency for Aquaculture Development (ANDA) was created in February 2011 to support *Halieutis*. ANDA is now leading the project to provide Morocco with 200,000 tons of aquaculture products by 2030.

[4]The Azur Plan was launched by the Ministry of Tourism in order to attract ten million tourists to Morocco by 2013. This number was later revised (Plan Azur 2020). The Azur Plan aims to creating six beach resorts along Morocco's coasts.

main purpose of governance in coastal zones is to promote sustainability. The way to insure sustainability is through Integrated Coastal Zone Management (ICZM).

15.3 Coastal Governance Is Based on Integrated Coastal Zone Management (ICZM)

ICZM is widely considered as the most adapted approach to coastal governance (Cicin-Sain and Knecht 1993). The ICZM is defined as: "a dynamic process for the sustainable management and use of coastal zones, taking into account at the same time the fragility of coastal ecosystems and landscapes, the diversity of activities and uses, their interactions, the maritime orientation of certain activities and uses and their impact on both the marine and land parts" (article 1, f, ICZM Protocol). ICZM means that when we consider the management of a coastal zone, we need to take into account the "whole picture"; for instance, we need to consider the interactions of the sea and the land (ecosystem approach), to adopt a bottom up approach (participatory approach), and to put together scientific, economic, social, legal, considerations (holistic approach). It is an integrated approach in the sense that it also encompasses both economic and ecological aspects. We cannot manage the coastal zone in a sustainable way if we do not preserve its ecosystems.

Therefore, coastal governance aims to acheive coastal sustainability. ICZM is enshrined in the Sustainable Development Goals (SDG), especially in SDG 14 (Neumann et al. 2017).[5] The 17 SDGs were adopted by the United Nations in 2015, as part of the 2030 Agenda for Sustainable Development. SDG 14 calls for the management of the coastal zone that takes into account the interaction of different activities, stakeholders, policies and their potential cumulative impacts. It is also apprehended under the bleu economy paradigm (CESE 2018). The concept of "Blue Economy" originates from the United Nations Conference on Sustainable Development held in Rio de Janeiro in 2012 and is also part of the UN 2030 Agenda. It aims at "the improvement of human well-being and social equity, while significantly reducing environmental risks and ecological scarcities". (UN 2014: 2) SDG 14 encompasses 10 targets. Target 14.2 aims to "sustainably manage and protect marine and coastal ecosystems to avoid significant adverse impacts, including by strengthening their resilience, and taking action for their restoration in order to achieve healthy and productive oceans".

ICZM is at the core of Morocco's Coastal Act, enacted in July 2015 (Idllalène 2016).[6] The Coastal Act is based on an ecosystems approach (articles 3 and 6),

---

[5] SDG 14 on "life below water," focuses on the sustainability of the oceans, including coastal areas. It encompasses ten specific targets. https://www.un.org/sustainabledevelopment/oceans/

[6] Morocco has also ratified the Madrid Protocol on Integrated Coastal Zone Management (ICZM Protocol) in 2012. The ICZM Protocol was adopted on 21 January 2008, and entered into force on 24 March 2011. The text of the Protocol is available at: https://paprac.org/iczm-protocol

planning (articles 4, 7, 8) and participation (article 9). According to this Act, ICZM is the "harmonious management of coastal areas which takes into consideration environmental, socio-economic and institutional aspects in order to guarantee balance and sustainability of the multiple functions of the coast" (Article 2–2).

The Coastal Act establishes two types of coastal planning tools: the National Coastal Plan (Plan National du Littoral (PNL)) and the Regional Coastal Schemes (Schéma Régional du Littoral (SRL)). The Department of Environment elaborates these plans "on the basis of the available socio-economic and environmental scientific data and by adopting an integrated management approach which takes into account the coastal ecosystem and climate change" (Articles 3, 7). The ecosystem-based approach includes both a spatial and temporal dimensions to the extent that it invites planners to plan for the long term, to factor-in the interrelationship of ecosystems, and to include all stakeholders in the management process (Forst 2009).

Before their adoption, the coastal planning documents have to be submitted to two committees: the National Commission for Integrated Coastal Management (Commission nationale de gestion intégrée du littoral (CNGIL)) and the Regional Consultative Commission (Commission régionale consultative (CRC). These commissions had only a consultative role, but their inputs are necessary before the adoption of the PNL and SRL (articles 5 and 9). While the PNL sets out the general orientations of coastal planning, SRLs are much more detailed. They designate the sites to be allocated to various economic, social and cultural activities according to an environmental diagnosis. They also establish precise zoning (areas to protect or develop, setbacks, no discharge zones, etc.). SRLs also applies both to the land and to the sea (article 8).[7]

While the Coastal Act entered into force in 2015. However, as it is underpinned by the PNL and SRLs, it is too early to analyze its impacts. The National Coastal Plan (PNL) was only validated by a decree published in the official Bulletin on June 2022.[8] This plan sets the general principles of coastal governance in Morocco and includes indicators in order to measure the enforcement of these principles (article 3).[9] Furthermore, so far only one coastal region (Rabat-Salé-Kénitra) has drafted a SRL, and this plan is now undergoing a process of consultation (Ministère de l'Energie, des Mines et de l'Environnement 2020). The SRLs of Tangier-Tetouan-

---

[7]For example, the SRL sets the use of areas subject to the plan, based on an economic, social, cultural and general environmental diagnosis of each of these areas. It sets measures for integrating marinas into natural sites and urban agglomerations. It suggests boundaries for the maritime areas intended for recreational vehicles, as well as the rules for using these vehicles and equipment. It sets additional measures necessary to ensure better conservation of the coast, including environmental awareness and education measures.

[8]According to the Coastal Law, a SRL must be compatible with the PNL. The PNL was validated by the Decree #2-21-965 of 16 Chaoual 1443 (17 May 2022) approving the National Coastal Plan, Official Bulletin #7096 of 2 Kana 1443 (2 June 2022), p. 719.

[9]Among these indicators are the "level of implementation of planning, development and management instruments, including sectoral, relating to coastal areas and assessment of their coherence and institutional coordination" (translation added).

Al Hoceima and Dakhla-Guelmim regions are currently being drafted and will need time for diagnostics before validation by a Decree (article 11).

ICZM in Morocco is thus a work-in-progress in its early stages. It is based on planning schemes initiated by the government. Regional administrations have an advisory role on how ICZM will be applied to their territories through an SRL. From this point of view, ICZM is challenging for coastal stakeholders, especially local authorities. These stakeholders lack clear jurisdiction over the coastal zone and share no common approach on how the coastal zone should be managed.[10]

15.4   Coastal Governance Is Challenging for Stakeholders

Morocco's Special Commission on a New Development Model's diagnostic report has shown that the weak vertical and horizontal convergence of public policies is a real obstacle to development (Royaume du Maroc, La Commission Spéciale sur le Modèle de Développement 2021). This is all the truer for coastal zones.

The national and local levels of coastal management are interconnected. Coastal stakeholders are involved in designing the SRL, which is the main ICZM tool, and are then responsible for its implementation. The diversity of stakeholders in coastal zones is explained by the very nature of areas that bring together governmental agencies operating on land (Department of Agriculture; the High Commission for Water and Forests, Department Equipment, etc.) and those operating at sea (Fisheries Department, Merchant Marine Department, Aquacultute Development National Agency, etc.). Because there is lack of clarity as to the legal attributions of each of these institutions, the diversity of stakeholders most often leads to an overlapping of jurisdictions and even to real conflicts.[11] These factors hinder the successful implementation of ICZM. In addition, as seen above, adoption of SRLs is slow.

• The national government. The scope and content of the government's jurisdiction over coastal zone does not favor the ICZM approach. Nor is it amenable to decentralization. *Territorial jurisdiction*: coastal stakeholders may intervene in the coastal zone simultaneously on land and off-shore. (Idllalène and Masski 2015). *Material jurisdiction*: coastal stakeholders can have double/redundant attributions. We might divide them into "pressure" and "conservation" stakeholders. For example, the Ministry of Fisheries and the Ministry of Equipment both intervene in the coastal zone and they both are pressure institutions (Sbai 2001). Because their attributions overlap, the pressure is exacerbated on the coastal zone. Another example of this is the "conflict" between the Ministry of Equipment and the Ministry of Forests; coastal dunes are simultaneously under the jurisdiction of both departments (Haut Commissariat des Eaux et Forêts).

---

[10] Yet according to the Coastal Act and to ICZM protocol ratified by the country, coordination and participation are the main basis for ICZM (article 7 and 14).

[11] See examples bellow.

- Local government. The local level of administration mirrors what happens at the national level. Three main aspects illustrate how ICZM is challenging for local authorities. Before delving into these aspects, we need to bear in mind that local governance has been strengthened by legal reforms brought about following the adoption of the 2011 Constitutional, which encourages the decentralization process (Bergh 2021: 491).[12]

In Morocco, the regions are the preeminent local authorities. They are in charge of designing economic development and land use planning strategies. The provinces/prefectures are at the intermediate level and have the role of promoting social development, particularly in rural areas, and strengthening cooperation between the municipalities ("communes") in their territory. Municipalities provide local services to citizens. The decentralization laws enacted in 2015 have expanded the jurisdictions local authorities (Badri 2019: 100). However, these reforms remain limited for coastal governance.

First of all, the attributions of local authorities, even if their application to the coastal zone (environment, urbanization, sand mining, etc.) is unequivocal, remain sectoral as they mainly relate the terrestrial part of the coastal zone. The extent of the local authorities' jurisdiction at sea is not clearly mentioned in the law. The legal texts *post* 2011 Constitution are silent in this regard. They include terminology such as: beach, environment, or even the word "littoral," but not in the same sense as in the Coastal Act.[13] For example, the Law No 14–113 on Municipalities mentions: "the management of the coastal zone situated on the territory of the commune, in compliance with applicable laws and regulations," yet does not refer to the coastal zone *per se*.[14]

Yet, according to Law No 14–113, municipalities have specific attributions that can be marshalled for coastal management purposes. The law enables them to: protect the environment, manage the coastal zone located under their territorial jurisdiction (in accordance with applicable laws and regulations), and management beaches, seashores, lakes and river banks located within their jurisdiction. The Organic Law of 2015 grants municipalities some share of jurisdiction coastal management, in addition to powers they already have with regard to beach

---

[12] Three main Statutes were enacted in 2015:

- The Organic Law #14.111 on Regions;
- The Organic Law #14.112 on Preefectures and Provinces;
- The Organic Law #14.113 on Municipalities.

The law on Regions was issued on July seventh, 2015 (promulgated by the Dahir #1-15-83) and published in the Official Bulletin #6440 of February 18, 2016, p. 197. The law on Communes/Municipalities #113-14 was promulgated by the Dahir #1-15-85 of 20 Ramadan 1436 (July, seventh, 2015) and published in the same Official Bulletin. The law on Provinces and Prefectures was also published in this official bulletin. Organic Laws have a Constitutional status which means that ordinary laws must comply with their provisions.

[13] See the definition of coastal zone above (footnote 1).

[14] Dahir #1-15-85 of 20 Ramadan 1436 (July, seventh, 2015).

management and environmental protection (Article 87). Thus, by applying the principle of subsidiarity, municipalities can benefit from the transfer of powers from the State in the field of the protection and restoration of historical monuments, cultural heritage and the preservation of natural sites (Article 90). The municipal council is competent in town planning, construction and land use planning (article 92). In addition, in certain cases the president of the municipal council can authorize occupation the public domain. However, under the Coastal Act, municipalities, and local authorities in general, have only an advisory role in the management of the coastal zone.[15]

The Coastal Act states that SRL must comply with the SRAT, which in principle may seem favorable to the harmonization of the role of the regions in local management. Regions, prefectures and provinces also have general attributions that can be used to manage the coastal zone (urbanization, environment protection, management of protected areas, etc.). Nine of Morocco's 12 regions have a coastline. 42 Prefectures and Provinces are coastal. (CESE 2022:13) The region is responsible for "the proper use of natural resources, their enhancement and preservation," as well as "contributing to the achievement of sustainable development" (Article 80) (Hamdaoui 2017). In fact, the region is competent for the development, implementation and monitoring of the regional development program and the regional land use planning scheme (Schéma Régional d'Aménagement du Territoire (SRAT)) (Article 81). The *walis* (viceroys of regions) and the governors of provinces and prefectures, all of whom represent the central government, are expected to coordinate public agencies of varying scales in order to promote ICZM. Nevertheless, if the extent of the jurisdictions of various local authorities is not clear to all, we cannot expect good local governance.

The Coastal Act is based on the participatory approach. This means that coastal governance should employ a bottom-up approach when designing plans and strategies. Yet the Coastal Act does not provide any tools that can enable local authorities proceed in this way. The predominant role of the national government in coastal management is at odds with the ongoing decentralization process, particularly in terms of "advanced regionalization." According to the Law on Regions, Regional authorities can replace the government under the principle of subsidiarity, particularly in the fields of energy, water and the environment (Article 94)[16] (Hamdaoui 2017). Moreover, according to the Act, coastal regional schemes should be realized in collaboration with local authorities. However, these authorities have only an

---

[15] Decree #2-15-769 of 15 December 2015 establishing the composition, number of members, the functions and operating procedures of the National Commission for integrated coastal management and regional commissions and terms of development of the national and regional coastal plans (Official Bulletin #6428 of 7 January 2016), p. 5.

[16] The Organic law on Regions, July seventh, 2015 (promulgated by the Dahir #1-15-83), Official Bulletin #6440 of February 18, 2016, p. 197. Under Article 95, when transferring powers from the state to the regions, principles of progressivity and differentiation between regions are taken into account. Moreover, this same article specifies that these transferred powers may be transformed into the regions' proper attributions, which will therefore require an amendment to Law #111-14.

advisory role. The law enables the Regions to ask the government to adopt the Coastal Regional Scheme (SRL), but does not allow them to decide on how coastal governance should be done. In addition, the government can launch the SRL even if the region does not request it (article 6).

For now, the SRLs do not exist; even though the Coastal Act dates to 2015. Without approved SRLs, that law remains inoperative. According to the Law, the SRL is not always based on administrative boundaries. To the contrary, the SRL should primarily conform to the configuration of coastal ecosystems. This means that local authorities from different regions may have to collaborate in order to set up a SRL.[17] Institutionalized collaborative tools exist within the new decentralization laws. However, these tools are difficult to apply (Badri 2019: 249). Another aspect of the limited powers of local authorities on the coastal zone is their feeble financial resources. In order to enable them to exercise their new attributions, local authorities can raise their own revenues to supplement funds transferred by the government. According to the Cour des Comptes (Court of Audit), the tax resources of municipalities cover on average only 54% of their operational expenses, the rest being provided by the Value Added Tax (VAT) transferred by the government (Zine El Alaoui 2017: 101). However, the transfer of jurisdiction from the government to the local authorities is not always followed by a transfer of corresponding funds. Local authorities struggle to fund basic public services such as sanitation, collection and treatment of waste and urban transport. The municipalities are the entities most involved in providing public goods, accounting or 78% of overall expenditure, while regions and provinces/prefectures use respectively 5% and 17% of national resources (Zine El Alaoui 2017: 101).

The issue of appropriate financial resources is raised in the Methodological Guide accompanying the SRL of Rabat-Salé-Kenitra which, as mentioned earlier, is in the consultation phase. This document suggests that "most regions that undertake SRL development must rely on direct allocations to their budgets," or on "reprogramming of resources within agencies or between government agencies" (Ministère de l'Energie, des Mines et de l'Environnement 2020: 6). They can also rely on alternative funding such as grants and donations from international organizations or foundations. Without proper funding, it is difficult for the regions to initiate a SRL, and even when they do initiate one, their role remains advisory only.

In sum, Morocco's legal framework for coastal governance is an interesting experiment in workable decentralization. Though only in initial stages of design and implementation, one can begin to glimpse just how it might actually work. The example of sand mining offers the opportunity to look into the inner workings of the new legislation.

---

[17] According to article 6 of the Coastal Act, neighboring regions can ask for a common coastal regional scheme when the coastal ecosystem extends beyond the administrative boundaries of any one of them. This is the case for example, when a wetland extends across two or more regions.

15.5 An Example of Unsustainable Coastal Governance: Sand Mining

Sand mining is a perfect illustration of Morocco's coastal governance vacuum. Sand is an essential commodity for modern construction, used particularly intensely for manufacturing concrete. Since 2019, the UNEP has been interested in the phenomenon of sand mining. Morocco is among the countries that are designated as victim of the anarchic exploitation of sea sand. The government is trying very hard to curb this scourge, but the numerous urban construction sites that abound throughout the country, make the lure of profit more attractive. Half of the sand in the country (ten million cubic meters a year) comes from illegal coastal sand extraction (UNEP 2019). Despite a succession of laws,[18] reports (CESE 2014a, b; Cour des Compte 2015, 2020), research studies (Mansoum 2016; Driffort 2021; Aangri et al. 2022), and journal articles (Mahmoud 2019),[19] neither local authorities nor the government are able to stop massive illegal sand mining (Deychillaoui 2019; EcoActu 2022; L'Opinion 2021). Sand mining reveals the imbrication of land-use and sea-use conflicts and the difficulties faced by the stakeholders who would manage them. It also reveals the lack of coordination between these stakeholders.

After examining the quantities extracted from 23 sand quarries near the city of Kenitra, the Court of Auditors (Cour des Comptes) announced that it appears that "the municipality did not take the necessary measures with the competent authorities to enforce compliance with the quantities authorized" (Cour des Comptes 2015: 101). In its latest report, the Regional Court of Auditors emphasized that the objectives of the Quarries Act were not achieved because of the non-operationalization of the quarries police, the wide divergence of control approaches adopted by provincial/prefectural commissions, and the lack of specific semi-annual control plans (EcoActu 2022). Therefore, a number of beaches are being eroded or will be threatened by erosion in the near future (UNEP 2019:7).[20] Besides, several other beaches are slated to disappear (Kasmi et al. 2020). Yet, these beaches are amongst the most attractive for domestic and foreign tourists. Some are even among the beaches designated to

---

[18] The Dahir of May 5, 1914 on the Exploitation of Quarries is the first legislation on sand mining. In 2002 another legislation was enacted (the law #08-01 on quarries exploitation) but never followed by implementation texts. This is why a Circular of the Prime Minister #6-2010 of June 14, 2010 relating to the exploitation and monitoring of quarries was adopted in order to fill the gap. In 2015, Law #27-13 on quarries was enacted (Dahir No 1-15-66 du 21 Chaabane 1436 (9 June 2015)). This law is followed by two decrees of application (Decree #2.17.369 of 11 Rabii I 1439 (30 November 2017) issued for the application of certain provisions of Law #27-13) and twelve ministerial decisions. 8 of these decisions were published on the same date, March 08, 2018. Other decisions followed, inducing the joint decision on dredging of 2020.

[19] Media always report illegal sand mining operations as run by a "Sand Mafia".

[20] According to the UNEP report of 2019, sand smugglers have transformed a large beach between Safi and Essaouira into a rocky landscape.

accommodate the beach resorts of the Plan Azur (UNEP 2019: 7; ARTE: 2021).[21] Economic losses engendered by illegal sand mining are estimated at five billion MD (Les écos 2017; Kasmi et al. 2020: 8).

The Quarries Act deploys a whole range of measures to counter the overexploitation of sand. First of all, an Environment Impact Assessment (EIA) is mandatory before opening a quarry (Article 13).[22] Second, the law establishes a regional planning system through regional quarrying schemes (Schémas Régionaux de Carrières). These schemes allow for monitoring materials extracted in every region according to the quantities allowed by the National Commission and the Provincial Commissions on quarry exploitation.[23] Finally, the law provides for measures of control and follow-up, notably by the institution of a quarries' police force and the adoption of administrative, civil and criminal penalties against quarry operators who do not comply with the law.

The Quarries Act reinforces measures to protect the environment, particularly in coastal areas. Thus, if during operations it turns out that the quarry harms the environment, specific measures can be undertaken, going as far as the closure of the quarry (article 24). In addition, the law establishes more restrictive measures for quarries on the seashore. In fact, the dredging area is strictly delimited at 20 m from the high-water line to avoid any disturbance of the marine ecosystem (Article 20). In addition, offshore dredging is only permissible 500 meters off the coast. Besides, a dredging quarry cannot operate more than 10 years. This duration is shortened to 12 months for test quarries (article 10-5).[24]

Despite these measures, coastal zones continue to be stripped of their most precious resource. This spoliation extends to the sea where the sand is dredged on the shore and disrupts marine ecosystems by hindering fishing activity. This

---

[21] This is the case for example of Larache beach. Illegal sand mining in this area is documented by journalists. It was the subject of ARTE TV documentary broadcast on January 22, 2021. According to the UNEP 2019 report "Asilah, in Northern Morocco, has suffered severe erosion of its beaches, due to regulatory issues, and pressures relating to tourism. Many of the structures near the coast are now in danger from erosion" (UNEP, p. 7).

[22] The Quarries Act refers to the IEA Act (Dahir #1.03.60 du 12 mai 2003, BO #5118 du 19 juin 2003). It also refers to the National Charter on Environment and Sustainable Development (CNEDD). Framework Law #12.99 on the National Charter on Environment and Sustainable Development, Official Bulletin #6240, 20 March 2014, p. 2496.

[23] Under Article 5 of the Law: The regional scheme for the management of quarries lays down the following: "1. A global strategic vision to ensure good quarry exploitation management to supply the market with quarry materials and to control the supply and demand as well as the cost of quarry materials.; 2. quarries exploited and abandoned, and reserves of exploitable materials; 3. the areas where sand mining is prohibited; 4. The objectives to be achieved in the rehabilitation of quarry sites on completion of exploitation.; 5. where applicable, the special conditions of exploitation applied to all quarries or to certain categories of quarries.; 6. Minimum width of areas of exploitation; 7. the areas where the materials may be mined by dredging; 8. the conditions of dumping of materials".

[24] Test quarries are intended for the extraction of a quantity of rocks whose volume does not exceed fifty cubic meters (50 m³), in order to recognize the nature and extent of these rocks as well as the conditions of their exploitation (article 1, Quarries' Act).

generates not only conflicts of use but also policy conflicts between coastal stake-holders. In fact, it confronts the policies of urban planning, environment, fisheries, tourism, forests, agriculture (UNEP 2019, 6)[25] and climate (UNEP 2019: 6; Idllalène and Van Cauwenbergh 2016:192).[26]

Coastal dunes take several years to form (Ranwell and Boar 1986: 9). They contribute to the sedimentary balance of beaches and protect them from erosion, infiltration of sea water and the impacts of rising sea levels. The Coastal Act prohibits the exploitation of maritime dunes (article 24). Yet it is these dunes that are coveted by sand mining companies. The Quarry Act imposes a quota of quan-tities of sand that can be mined, but these quantities are largely exceeded. It takes an hour for shovels to destroy the biodiversity of the maritime dunes that took years to build up. Companies leave after mining beaches, doing nothing to restore them as required by law (Chapter VI, Quarries Act). The landscape left by sand mining companies is a picture of desolation. All of this happens in full view of local authorities. The few visits by the provincial commission are doomed to failure because the sand companies are so well organized that they simply hide most of their trucks.[27] Even if the law gives them a role in the control of sand quarries, the local authorities often lack awareness (because they wrongly think that the sand will renew itself rather quickly) and leave hectares of sandy beaches within the reach of businesses that loot them all day long and into the night (which is in violation of the law). Often the roads that serve the sand mining areas are deliberately left in a degraded state, so as to make it difficult to control the activity. Yet, these roads are under the jurisdiction of municipalities.

Illegal sand mining has economic and human impacts even outside the exploita-tion sites since the buildings constructed with this sand do not conform to minimum safety conditions. Every year, several buildings collapse on their inhabitants due to anomalies in their foundations.[28] The Department of Urban Planning passed a law in order to strengthen the penalties against construction companies (Law 66-12 relating to the control and repression of offenses in matters of town planning and construc-tion). But these construction companies are not necessarily those who conduct sand mining. Trade in construction sand is opaque. Construction companies look for the

---

[25] Agriculture lands can be affected by river erosion and by the lowering of water table.

[26] Climate policy is concerned by sea level rise and flooding. The UNEP 2019 report on Sand and Sustainability details these impacts by emphasizing that flooding and storm surges can affect houses and infrastructure. The increase in bed load or channel shortening can cause downstream erosion, including the undercutting or undermining of engineering structures such as bridges, side protection walls and structures for water supply.

[27] Local survey in the south of Safi, June 2021.

[28] Driffort reports the testimony of the geologist Youssef Zerhouni who explains that the unregulated use of sand worsens the obsolescence of constructions. "Very often, the sand is neither washed nor sorted, and it is therefore common that it contains salt and organic matter" (translation added). This is corroborated by the aforementioned UNEP report (2019) which states that half of the sand marketed in Morocco (or ten million cubic meters per year) is illegally extracted.

cheapest sand. Little attention is paid to its origin or content. After all sand looks alike! (ARTE 2021).

Sand mining lies at the intersection of several preoccupations (legal, socio-economic, political, environmental...). It affects numerous economic sectors: construction, tourism, agriculture, forests. It all takes place in a very fragile environment and in a very narrow space, the beach. Consequently, there are huge governance challenges.

The law requires integrated coastal zone management. For example, the Regional Quarries Schemes allows official agencies and local authorities to put together a holistic vision for quarries so as to avoid unbridled sand mining. However, in the field this is not happening. The Ministry of Tourism cannot create beach resorts in areas were the sand has been removed. At the same time, this agency needs more and more sand to build beach resorts. The Ministry of Fisheries has recently denounced the fact that the Minister of Equipment allowed dredging in very sensitive coastal areas (Kabbaj 2021). The Department of Urbanism denounces the infringement of urban regulations. The problem is exacerbated by the lack of financial resources and legal empowerment at the municipal level. In fact, Municipalities are not responsible for what happens at sea because the sea is under the jurisdiction of the national government. If national government agencies do not act because the power of the cement lobby, what chance does a municipal employee—or even an elected mayor—have of protecting coastal dunes from being stripped? The majority of sand quarries are located in rural areas, where municipalities lack essential resources and illiteracy is widespread.

Sand mining is a structural issue that reflects the complexity of coastal management and the limits of local governance. Its solution will depend on how public authorities juggle these two imperatives. These solutions revolve around good governance, which the Moroccan Constitution of 2011 and subsequent decentralization laws are supposed to promote.

15.6   Conclusion

Coastal Zone Law is based on the principle of ICZM. This principle supposes coordination among public agencies and the participation of local authorities in decision-making. Morocco's coastal zones encapsulate the contradictions of decentralization in Morocco. Public participation is lacking and local governance is limited. Besides, public agencies struggle to find a common vision for territorial governance in these very sensitive and yet rich territories. Therefore, there is a gap between coastal management principles as agreed upon in the law and coastal management on the ground.[29] Sand mining perfectly illustrates this conundrum. The limits of local governance appear clearly through the feeble enforcement of the

---

[29] In the Coastal Act of 2015 but also in the ICZM Protocol ratified by Morocco in 2012.

law (Quarries Act, Coastal Act, Environment Impact Assessment Act, National Charter on the Environment and Sustainable Development).

Consequently, it is important in the current Moroccan political context, which is well disposed towards local governance (SDGs, Blue economy, advanced regionalization, New Development Model...), to take this opportunity to clarify the responsibilities of local authorities in the coastal zone and give them the means to manage it more effectively. One inspiring example for this reform would be the National Association of Coastal Municipalities in France (Association national des élus du littoral, (ANEL)). ANEL was created in 1978. It has since brought together elected officials from coastal municipalities around the specific issues of economic development and the protection of the coastal and marine areas.[30] However, the experience of ANEL should be taken carefully as we need to take into account the specificities of local elected officials, particularly in rural areas of Morocco, who often lack the necessary means and skills to participate in such a network. In fact, about half of elected municipal officials have a level of education that does not exceed the primary level. Paradoxically, under the new reform, no level of education is required to lead a local executive (Badri 2019: 107).

In the process, it is also important to clarify the interrelationships of local authorities with the attributions of central authorities in the coastal zone. In effect, coastal governance first involves accelerating the process of adopting regional coastal schemes and strengthening the capacities of local actors (local authorities and associations) within the regional coastal commissions in order to ensure the adaptive management of this space. Recently, while this chapter was in its publication process, the Conseil Économique, Social et Environnemental (CESE) released an Opinion titled "Quelle dynamique urbaine pour un aménagement durable du littoral ?" ("What urban dynamics for sustainable development of the coastal zone?") (CESE 2022). In its 37-page report, the CESE draws up a provisional assessment of the implementation of the Coastal Act, pointing out the flaws that hinder this implementation (notably the weak coordination of public agencie, etc.). It recommends, as a solution, setting up a strong agency for coastal zone governance. This proposal is reminiscent of the old version of the Coastal bill, which encompassed provisions related to the creation of a Coastal Agency[31] (Idllalène 2016: 184).

But the Agency suggested by the CESE is of a different order. The CESE calls for the creation of a coastal agency based on the model of Marchica Lagoon Agency in Nador (CESE 2022, 8, 28). The Marchica Agency has strong powers that has allowed it to intervene in the coastal landscape of the Nador city. Based on its

---

[30] The ANEL website is https://anel.asso.fr/page-d-exemple-2/

[31] However, the Agency provisions were withdrawn from the last version of the Coastal bill in order to facilitate its adoption as it was considered that the Agency would duplicate the work of the other coastal stakeholders. Besides, the agency has a cost that will have to be paid (by whom?). It is noteworthy that the draft coastal law remained in the legislative process for more than one decade.

own special management plan, the Marchica Agency's founding legal text allows it to establish rules that can over-ride land tenure rules, the Decentralization framework, and to other legal texts applicable to the coastal zone.[32] For example, in order to issue its management plan, all real estate transactions related to properties (registered or not) and sites within the development area were suspended, except for those specifically concluded with the Agency for the development of the Marchica Lagoon.[33] Chaired by the Prime Minister, the Board of Directors of the Marchica Lagoon Agency issued the development plan for the lagoon which was approved by a decree at the behest of the Minister of the Interior.[34] Morocco has established a few other Agencies on the model of the Marchica Lagoon Agency. Among these are the Bou Regreg Agency in Rabat-Salé. The urban development along the Bou Regreg Estuary which has ensued can be considered "a mega-project where local powers are suspended in the name of greater efficiency" (Dauphins and Sperandio 2015: 8).[35]

Should the CESE recommendation be followed? How does the creation of special agencies with extraordinary powers to circumvent laws square with Morocco's official policy orientations: decentralization, empowering local government, integrated management? Will not the return to the Agency's model further weaken the role of local authorities in coastal governance? Yet even the CESE, in its Opinion, reiterates the importance of "granting municipalities, in accordance with the principles of local democracy and decentralization, decision-making prerogatives in terms of the development of their territory, urban planning and the preparation of urban planning documents" (CESE 2022: 8).

References[36]

Aangri A, Hakkou M, Krien Y, Benmohammadi A (2022) Predicting shoreline change for the Agadir and Taghazout coasts (Morocco). J Coast Res 38(5):937–950

Amine R (2016, 22 Jan) Les clusters au Maroc: vers l'émergence d'une nouvelle politique industrielle territoriale. Marché et organisations 26(2):93–120

---

[32] The Dahir creating Marchica Lagoon Agency was issued before the Coastal Act (issued in 2015). Nevertheless, even before the publication of this Act, the coastal zone was regulated by disparate legal texts.

[33] Article 4 of the Dahir #1-10-144 of 16 July 2010.

[34] According to the Dahir #1-10-144 of 3 Chaabane 1431 (16 July 2010) promulgating the law #25-10 relating to the development and enhancement of the Marchica Lagoon site (Official Bulletin #5862 of 23 Chaabane 1431 (5 August 2010), p. 1522) the Marchica Lagoon Agency consults with local authorities and other coastal stakeholders before establishing its special development plan. However, these authorities only have an advisory role and have to formulate their opinion within 1 month after the transmission of the drafted special plan.

[35] This is the case for example of the Bouregreg Agency also triggering the management of the coastal zone of Rabat-Salé.

[36] All websites cited were accessed in September 2022.

ARTE Reportage (2021) Maroc, Razia sur le sable. https://campus.arte.tv/sequence/la-mafia-du-sable-defigure-les-plages-ma-1

Badri L (2019) La décentralisation au Maroc: quelles perspectives pour la gouvernance locale et le développement territorial? (Cas de la régionalisation avancée). Dissertation. Université Grenble Alpes

Bergh SI (2021) Democratic decentralization and local development: insights from Moroccos advanced regionalization process. In: Crawford G, Abdulai A (eds) Research Handbook on Democracy and Development. Edward Elgar Publishing, Cheltenham, pp 482–501

Cicin-Sain B, Knecht RW (1993) Implications of the earth summit for ocean and coastal governance. Ocean Development & International Law 24(4):323–353

Cour des Comptes Maroc (2020) Rapport financier annuel publié sur le site. https://www.courdescomptes.ma/publications/

Dauphins E, Sperandio C (2015) L'urbanisme au Maroc face à son environnement institutionnel. https://urbanistesdumonde.com/en/

Deychillaoui M (2019) Surexploitation des carrières de sable: Amara saisit la justice. https://fr.le360.ma/societe/surexploitation-des-carrieres-de-sable-amara-saisit-la-justice-195061

Driffort D (2021) L'extraction du sable au Maroc, de la ressource au produit. Les Cahiers de la recherche architecturale urbaine et paysagère. [Online]. http://journals.openedition.org/craup/7464

Eco Actu (2022) Region Casablanca-Settat: les infractions dans la gestion de l'exploitation du sable perdurent https://www.ecoactu.ma/gestion-exploitation-des-carrieres/Cour Regionale des Comptes, Casablanca-Settat, 2019–2020

Forst MF (2009) The convergence of integrated coastal zone management and the ecosystems approach. Ocean Coast Manag 52(6):294–306

Hamdaoui S (2017) La régionalisation avancée au Maroc: entre la lutte contre les changements climatiques et la protection de l'environnement. Revue juridique de l'environnement 42(3):425–442

Idllalène S (2016) Integrated coastal zone management in morocco: from improvised norms to formal law. Int J Mar Coast Law 31:168–188

Idllalène S, Masski H (2015) Les aires marines protégées: nouvel outil de gouvernance côtière? Le cas du Maroc. In: Bonnin M, Laé R, Behnasi M (eds) Aires Marines Protégées Ouest Africaines, défis scientifiques et enjeux sociétaux. IRD, Paris

Idllalène S, Van Cauwenbergh N (2016) Improving legal grounds to reduce vulnerability to coastal flooding in Morocco: a plea for an integrated approach to adaptation and mitigation. Ocean & Coastal Management 120:189–197

Kabbaj M (2021) Akhannouch a eu gain de cause du dragage du sable marin à Larache, Maroc Hebdo. https://www.maroc-hebdo.press.ma/akhannouch-gain-cause

Kasmi S, Snoussi M, Khalfaoui O, Aitali R, Flayou L (2020) Increasing pressures, eroding beaches and climate change in Morocco. J Afr Earth Sci 164:103796

L'Opinion (2021) Province de Larache. Du contrôle de la gestion des carrières de sables. https://www.lopinion.ma/Province-de-Larache-Du-controle-de-la-gestion-des-carrieres-de-sables_a12368.html

Laouina M (2016) Gestion en concertation de l'espace littoral marocain, les contraintes et les perspectives. In: Mansoum M, Benali A (eds) Les littoraux marocains: changement climatiques et stratégies de gestion, Paysages géographiques, vol 2, pp 11–30

Les écos (2017) Les nouvelles modalités d'exploitation détaillées, https://leseco.ma/business/les-nouvelles-modalites-d-exploitation-detaillees.html

Lmariouh A, Askour K (2021) L'action publique au Maroc face à l'incohérence temporelle: Des territoires sans prospective. International Social Sciences and Management Journal (ISSMJ) 5:86–105

Mahmoud M (2019, 13 Mai) Les plages du Maroc dépouillées par des mafias du sable, TelQuel

Mansoum M (2016) La gestion de l'érosion côtière au Maroc. In: Mansoum M, Benali A (eds) Les littoraux marocains: Changement climatiques et stratégies de gestion, Paysages géographiques, vol 2, pp 31–62

Nakhli S (2010) Pressions environnementales et nouvelles stratégies de gestion sur le littoral marocain. Méditerranée 115:31–42

Neumann B, Ott K, Kenchington R (2017) Strong sustainability in coastal areas: a conceptual interpretation of SDG 14. Sustain Sci 12:1019–1035

PACC (2022) Le développement durable du littoral et des espaces côtiers. Projet de jumelage Maroc-UE pour la convergence réglementaire environnementale, Programme d'Appui à la Compétitivité et à la Croissance verte, financé par l'Union européenne. https://competitivite-pacc.ma/wp-content/uploads/2022/03/Sensibilisation-protection-du-Littoral-PACC-MAROC-UE-2022-1.pdf

Ranwell DS, Boar R (1986) Coastal dune management guide. Institut of Terrestrial Management. School of Biological Sciences, University of East Anglia Norwich

Royaume du Maroc, CESE (2014a) Avis du Conseil Économique, Social et Environnemental. Projet de loi n°81-12 relative au littoral. Saisine No 13/2014

Royaume du Maroc, CESE (2014b) Avis du Conseil Économique, Social et Environnemental. Projet de loi n°27-13 relative à l'exploitation des carrières. Saisine No 12/2014

Royaume du Maroc, CESE (2018) L'économie bleue: pilier d'un nouveau modèle de développement du Maroc, Rapport du Conseil Économique, Sociale et Environnemental, Auto-Sasine n° 38/2018, www.cese.ma

Royaume du Maroc, CESE (2022) Avis du Conseil Économique, Social et Environnemental. Quelle dynamique urbaine pour un aménagement durable du littoral? Auto-Saisine

Royaume du Maroc, Commission Spéciale sur le Modèle de Développement (2021) Le nouveau Modèle de Développement. Libérer les énergies et restaurer la confiance pour accélérer la marche vers le progrès et la prospérité pour tous. Rapport général

Royaume du Maroc, Cour des comptes (2015) Rapport sur la fiscalité locale

Royaume du Maroc, Ministère de l'Energie, des Mines et de l'Environnement, département de l'Environnement, Groupe Banque Mondiale, Ministerio de Ambiante y de la Tutela del Territorio y del Mar, Projet d'Assistance Technique (2020) Integration de l'approche de gestion intégrée des zones côtières dans les processus de planification et de développement territorial, Elaboration du Schéma Régional du Littoral (SRL) de la Région Rabat-Salé-Kénitra (RSK), Guide méthodologique pour l'élaboration du Schéma Régional du Littoral (SRL)

UNEP (2019) Sand and sustainability: Finding new solutions for environmental governance of global sand resources. http://www.unepgrid.ch/storage/app/media/documents/Sand_and_sustainability_UNEP_2019.pdf

United Nations (2014) Blue Economy Concept Paper. https://sustaina-bledevelopment.un.org/concent/documents/2978BEconcept.pdf

Zine El Alaoui S (2017) La décentralisation entraine-t-elle des comportements stratégiques des collectivités locales au Maroc? Revue d'économie du développement 25(2):95–114

Chapter 16
Some Conclusions and Reflections

Robert Home

This author was born into the aftermath of the Second World War when a new international political settlement was being framed around the newly-formed United Nations. Its first General Assembly, then comprising 51 nations but its numbers increasing four-fold in the following decades, was held at Westminster Central Hall in London in January 1946, with Britain's overseas colonial empire beginning to fragment along with other empires. Many colonial administrations became sovereign nation-states with their territorial boundaries protected by the *uti possedetis* principle, but in the succeeding decades the new nation-state dispensations sometimes proved to be unstable. This was especially so in the Middle East, where the international recognition of the state of Israel, the geopolitics of oil and gas, and the arbitrary boundary lines between French and British spheres of influence drawn by the Sykes-Picot agreement of 1916 led to international disputes and political upheavals, most recently since the 2003 invasion of Iraq and the civil war in Syria that began in 2011. Nation-building projects aim to unify the people living within the state so that it can become and remain politically stable and viable, with legitimate authority conferred through elections, constitutions and majority rule. Central governments co-exist with various institutions of local governance and have had to accommodate multiple local non-state actors.

This book brings together contributions from across Africa and parts of the Middle East. It does not maintain a distinction between 'sub-Saharan' and North Africa, acknowledging that such a separation is artificial, a historical construct with colonial and religious origins that obscures much that is similar and sees the great desert as a significant barrier to social interactions.

The unstable dynamic between national and local governance institutions is a recurrent theme in the collection. A case study of Tunisia, in Mohamed El-Mensi's

R. Home (✉)
Department of Business and Law, Anglia Ruskin University, Cambridge, UK

K. Darmame, E. Ross (eds.), *Local Governance and Development in Africa and the Middle East*, Local and Urban Governance,
https://doi.org/10.1007/978-3-031-60657-1_16

Chap. 6, shows how the French protectorate regime sought to reduce the power of the local clans and tribes by creating communes (municipalities) with defined boundaries supervised by colonial administrators. The change from colonial protectorate to an independent nation-state in 1956 involved an exercise in structuring a new national identity through a centralized state and inherently authoritarian public administration, prioritizing security and control of the society over local development concerns. Subsequent constitutions and laws provided for various degrees of administrative decentralization to the communes, but they remained under-resourced and struggled with their role as agents of local development. The 2014 constitution that followed the Arab spring and change of regime provided for greater local autonomy, yet local governments remained subordinate to the central state's administrative apparatus. While some positive outcomes emerged from this new legal framework, the subsequent 2022 constitution reversed some of the changes and restored greater central control.

Three chapters present further case studies, from Jordan and Morocco, of the modernising and centralising trends of the post-colonial period, when supposedly a-political but powerful professional actors introduced grand modernist concepts of town and regional planning, particularly master plans for metropolitan areas, into government. Fuad Malkawi's Chap. 8, 'Ideology and Power: Defining the Greater Amman Metropolitan Area,' shows how the Jordanian government, transitioning from British colonial informal rule, transferred control of the rapidly growing city of Amman from elected local authorities to appointed and unelected planners. Town planning practices, modelled upon British reconstruction experience after the Second World War, were presented as technical solutions to a range of urban development problems, but allowed planning experts to over-ride municipal institutions, and largely excluded public participation. Another case study in Chap. 9, 'The Endless Challenge of Local Governance in Casablanca', by Khadija Darmame, Abdelkader Kaioua and Eric Ross, investigated the planning arrangements for Morocco's biggest metropolitan area. More than 50 years of administrative and territorial reform have not prioritized equitable or inclusive growth. Instead, they and created a governance deficit in addressing significant urban development challenges. Various stakeholders have been inhibited from carrying out their responsibilities, and public administrative agencies are entangled in power relationships and cross-purpose decision-making. Prerogatives overlap, actions are needlessly duplicated, and policies poorly implemented at the local level. This governance deficit is exacerbated because those elected bodies most representative of popular will are the least empowered in relation to the centralized state agencies. If Greater Casablanca is to cope with urban growth and ensure effective local participation in decision-making, new forms of urban governance will be needed which can enhance innovation and collaboration between all actors and sectors. Another contribution on Moroccan territorial and economic dynamics, Chap. 4 by El Hassan Farhat and Khadija Darmame, seeks to highlight the potential of scientific research and technopoles in fostering inclusive regional development through policies of advanced decentralization that can institutionally couple the outputs of research institutes and universities, particularly high-technology industries, to the needs of local economic actors.

All these governance experiments took place in a context of rapid population and urban growth, often accompanied by large-scale forced displacement of people either because of right of imminent domain, political conflicts or environmental changes. Increased international attention to basic security introduced related issues of humanitarian action, human rights and democratic participation, and these have often become rallying points for citizens to hold their governments to account in promoting broad human development. Several chapters in the collection address such issues.

Ngang and Dick-Sagoe's Chap. 2, on the right to development in African governance, offers a legal perspective, recognizing the significance of the African Charter on Human and Peoples' Rights (the 1981 Banjul Charter) in placing obligations upon national governments in Africa to ensure socio-economic and cultural development for their peoples. While the right to government has been a contested idea, the Banjul Charter was soon followed by the United Nations Declaration on the Right to Development in 1986. African governments have, however, often failed to create mechanisms—the legal and administrative procedures, organizational entities able to ensure transparency, accountability and citizen participation—for the right to development to become effective, as the authors examine in two specific cases. Oku white honey production in a mountainous region of Cameroon was the first registration in Africa of 'protected geographical indication' (from 2013). It aimed to ensure standards for production, but there remains limited public awareness in application and practical protection. In the second case, the Maasai communities of Kenya and Tanzania were largely deprived of their ancestral lands by the colonial powers, and subsequent government promotion of nature conservation and wildlife parks to attract tourism revenues has denied the Maasai much of the potential benefits within their rural communities.

Within the human rights policy discourse, international agencies have increasingly sought to improve equality and rights for women, and Safaa Monqid's Chap. 12 addresses issues of women's involvement with urban governance in the Arab world. In the twenty-first century urban women have adopted innovative approaches to improve their participation in public life, and particularly to reclaim and transform public spaces in their cities. Women have always played an essential role in family, community, local and national organization, and their collective action can be important even if informal. The segregation of urban space between public and private/domestic, the first mainly a domain for men, and the other for women, is now increasingly challenged by informal women's actions, though these are resisted by city administrations.

Growing population pressures are translating into demand for housing and livelihoods in the rapidly expanding informal areas around the state-planned urban spaces formed by colonialism. Colonial governance structures and urban formations have largely continued under the political settlements at independence, and create a gulf between the institutions, language and cultures of governance on the one hand, and the mass of people struggling with basic issues of daily survival on the other. Government institutions do not have a monopoly over public action, and a multitude of actors with capacity for collective action are pushing to be accommodated.

Government failure to provide or facilitate public goods to informal settlements under conditions of rapid urbanization is addressed in Danso-Wiredu's Chap. 11, on access to housing, water and sanitation in four poor neighborhoods in Ghana. The national and local government institutions have failed to provide these to low-income citizens, and non-state actors, such as traditional courts and authorities, and local voluntary associations especially, have enabled a measure of local self-government in these communities. They rely upon social capital and local citizen and stakeholder participation, but remain handicapped by poverty. Thus the state still has a role in enacting policies that can improve the housing and community infrastructure in poor communities.

The guardianship of public monies against corruption is becoming increasingly important. Akono Olinga's Chap. 5 addresses issues of financial transparency in Francophone sub-Saharan Africa if sustainable development is to be successfully pursued. Since 2010 the states of Cameroon, Gabon, Burkina Faso and Côte d'Ivoire have all enacted national laws for greater transparency in public finance, linked to the consolidation of local democracy and local sustainable development. As local authorities have acquired increased powers, they have become increasingly politicized. The author advocates monitoring systems for local public policies, and dynamic, flexible controls over local public finance that recognize local social and geographical realities and can be adapted over time. Political control of local public finances requires greater citizens' involvement and control of local governance through improved participatory democracy.

Rey and Petitpierre's Chap. 10, on international standards for urban development projects in Guinea, investigates similar issues of protecting citizens' human rights. International norms and standards on governance, particularly the Equator Principles on environmental and social impacts of development projects, can effectively promote good local governance, but adherence to the Equator Principles depends upon the willingness of the funding institutions to enforce them. In a study of urban infrastructure projects in Conakry, they found that the government signed accords with the People's Republic of China without applying the Equator Principles or the World Bank's best governance practices, such as stakeholder participation and impact assessments. The government undertook major road projects and a new government complex, taking land without compensation or consultation, and demolishing people's homes and businesses without due process. Meanwhile, delays in completing the road projects caused massive traffic congestion, again with no citizen consultation.

Environmental issues and resource exploitation has been another area of contestation explored in several chapters in the collection. Zeina Moneer's Chap. 7, on environmental activism explored the rise, after the so-called 'Arab Spring', of numerous environmental movements, particularly in Algeria, Lebanon, and Morocco. Such environmental activism was concerned with more than environmental issues, but reflected everyday struggles to secure wider political and economic rights. Chapter 14, by Karen Rignall, Atman Aoui and Moulay Ahmed el Amrani, investigates what they call 'extractivism' in mining and the local dynamics of territorial governance in Southeast Morocco. Contrary to its stated intention,

Morocco's 2015 mining code continues colonial-era wealth extraction policies. The central government maintains control of granting permits for exploration, exploitation, compliance, and decommissioning, leaving contradictions in jurisdiction that complicate territorial governance for communes trying to plan comprehensively for economic development. Relations between residents and local elected officials on the one hand, and the state and private companies on the other, are heavily weighted for the latter's benefit. The 2015 law allows a lack of transparency at every stage of the extraction process, including disclosure of revenue flow and mitigation of local impacts. Attempts at redress provoke repressive measures. Better public knowledge about the legal framework for mineral extraction might offer some tools for local actors, but local communities may not be able to challenge the process. Similar local issues were discovered in relation to two apparently different types of large-scale project, for copper mining and for solar energy generation.

An even more basic resource—water—is increasingly scarce in the Middle East and parts of Africa. Ahmed Haj Asaad and Khadija Darmame's Chap. 13, on governance of water access in the Syrian Jazeera region, presented a case study of the Al Ayn Springs (not to be confused with the World Heritage site of Al-Ain in the United Arab Emirates). Ground-water depletion and drying up of many watercourses have contributed to water insecurity, with the issue being weaponized by powerful interests since the outbreak of the Syrian conflict in 2011. The region comprised different ethnic and religious communities of Kurds, Arabs, Assyrians, Armenians, and Yazidis which had co-existed for centuries. Poor, uncoordinated, water extraction practices on both sides of the Syrian-Turkish border was degrading groundwater resources even before the conflict started. Since then, fragile intercommunal social relationships have been further disrupted, and access to water supplies has been politized and even weaponized. Yet, despite this, if viewed beyond narrow professional engineering perspectives, water management has the potential to foster social cohesion. It remains to be seen if the multiple actors involved in the region can devise an appropriate system to achieve this.

Samira Idllalène's Chap. 15, on local governance along Morocco's coasts, investigates attempts to apply the concept of coastal zone management since the enactment of the Coastal Act in 2015. This act subjects coastal zones to an integrated management policy, following international norms and principles, and seeks to establish a two-tier (national and regional) planning regime. Among the many issues affecting coastal zones, sand procurement for the construction industry has become a lucrative activity for local entrepreneurs in coastal areas, and is being pursued with few effective controls. Coastal municipalities, especially rural ones where most sand mining occurs, lack the financial resources, expertise and legal powers necessary to manage their coastlines, while the sea remains under the jurisdiction of the national government. Ad hoc national agencies with wide powers have been created, notably the Marchica Lagoon Agency in Nador, and the Bouregreg Agency in Rabat-Salé, but these contravene Morocco's official policy orientations promoting decentralization, the empowerment of local government, stakeholder and citizen participation, and integrated management of territories and resources.

After much consultation, in 2015 the international community agreed the Sustainable Development Agenda 2030, with its 17 goals and many accompanying targets and indicators to be met. Home's Chap. 3, on urban governance and climate action challenges in Africa, explores the implications of this for local development and governance institutions, with particular attention to SDG 11 (sustainable cities) and SDG 13 (climate action). With territory and people arguably the two greatest basic resource in a country, over-arching issues of land governance are becoming crucial, at international, national, local and community levels, and were given recognition in the Arab Land Initiative' of 2020–23. A new cross-disciplinary field of policy and research has emerged concerned with geographies of governance, exploring the different types and levels of territorial governance and the legal reforms necessary to make good land governance. Central governments have sought to keep control and design the structures of local governance, but increasing challenges need more local action and support for local governance.

These conclusions are being written in October 2023, during a year of extreme weather events in many parts of the world, including in regions studied in this volume. Extreme heat, floods, drought and water shortage—as well as political disruptions and widespread displacements of people—continue as we write. All of these require the concerted actions of both central and local governance institutions, yet all too often the people and communities most directly affected have to cope for themselves, without the help they need from public emergency services. The flooding of the Libyan town of Derna cruelly exposed the failings of these public services, and the consequences of a nation-state divided by conflict into two separate administrations. Attempts to over-ride tribal and clan loyalties through modern forms of territorial governance still have to adapt to local needs and accommodate local actors. The global human population has more than trebled since the Second World War, with drastic impacts upon climate, eco-systems and governance. We are in a new era of anxiety and awareness of existential threats, especially climate change. Now, in the words of the American Benjamin Franklin, writing before the American War of Independence: 'We must, indeed, all hang together or, most assuredly, we shall all hang separately'.

3 October 2023   Robert Home

Index

The manufacturer's authorised representative in the EU is Springer
Nature Customer Service Centre GmbH, Europaplatz 3, 69115 Heidelberg,
Germany. If you have any concerns regarding our products, please
contact ProductSafety@springernature.com

Printed and bound by CPI Group (UK) Ltd, Croydon, CR0 4YY
27/04/2026
02097573-0006